SHIELDED Metal Arc Welding

Theories and Techniques

By: William L. Ballis, P.E.

Shielded Metal Arc Welding
Theories and Techniques
by William L. Ballis, P.E.

Printed in the United States of America

ISBN 9781612153995

www.xulonpress.com

This book is dedicated to my late loving wife, Wanda;

To the Columbia Gas Company welders;

To my late friend and classmate, Eldon Brandon; and

To my late friend Kenton Ridge High School industrial arts teacher John Mills

Preface

The purpose of this book is to teach students how to use the Columbia Gas "puddle" welding method. I have used this method to teach the adult welding classes at the Tolles Career and Technical Center in Plain City Ohio for 30 years. After completing the 24-hour shielded arc welding course, most of the students can pass the AWS D1.1 structural welder qualification test.

The puddle method is an easy, fast, economical way to train welders. The heart of the method is reading the puddle (liquid weld pool) to make continuous corrections of welding variables during the welding operation. Some of the variables controlled are travel speed, work angle, travel angle, arc length, complete fusion to both bevels, correct welding current, correct polarity, and correct welding machine set-up and operation.

The first seven chapters describe the techniques used in the adult welding classes at Tolles. Chapters 8-14 describe pipeline welding techniques used by Columbia Gas for over 60 years.

The third edition includes the addition of an appendix that describes the 17 principles of success that Andrew Carnegie used to accomplish his goal of replacing cast iron with steel at a lower price. Mr. Carnegie shared these principles with Napoleon Hill in 1908. I have taught these 17 principles to the inmates at the Madison Correctional Institution in London, Ohio and the Orient Reception Center in Orient, Ohio. I strongly believe in the power of these 17 principles. If one takes these principles to heart, he or she will be inspired to live their life to the fullest and obtain personal achievement in all areas of their lives.

William L. Ballis, P.E.
wballis@ee.net
Mechanicsburg, Ohio

Acknowledgments

T he author wishes to thank Cecil DeHaven for using his many years of welding experience to teach the author how to weld. I will be forever grateful for his patience and guidance in showing me the hows and whys of manual welding.

The American Welding Society provided helpful illustrations and encouragement. Some other organizations who provided important information were H.C. Price Company, Lincoln Electric Company, American Gas Association, Columbia Gas System, T.D. Williamson, Inc., and Kerr Engineered Sales Company.

The author wishes to thank Dr. Sudheer Pimputkar at Battelle Memorial Institute for his review and encouragement. Also, Ohio Governor Voinovich for his support of job training skills and answering my letter concerning the upgrading of Ohio's vocational welding instructors.

Last but not least, many thanks to Joe Stets and my wife Wanda for valued comments, support and editing.

About the Author

In 1964 the author graduated from The Ohio State University with a Bachelor's degree in Welding Engineering. He has served as Manager of Materials, Standards, and Testing for Columbia Gas Distribution Companies for twenty-three years. He has six patents and is the sole inventor of an Automatic Pipe Welding Machine. He has taught Adult Fundamentals of Welding for over twenty years at the Tolles Technical Center, Plain City, Ohio. He has been a member of the American Welding Society for thirty-five years and served on the following technical committees: AWS D10 Piping and Tubing, AWS Technical Activities Committee, ASME B16 Valve Standards Committee, American Petroleum Institute Standard 1104, Welding of Pipelines and related facilities and AWS B2.1 Welding Qualification and American Gas Association Plastic Materials Committee. He revised the Oxyfuel Gas Welding Chapter in Volume 6 of the American Society of Metals Welding, Brazing, and Soldering handbook.

The author has written papers on Oxyacetylene Pipe Welding and Arc Welding of Pipe for the AWS Welding Journal and has given AWS presentations on "Welder Training", and "Pipeline Welding Processes as a Welder Sees Them", at the following cities: Dayton, Ohio; Columbus, Ohio; Washington, D.C.; Cleveland, Ohio; Indianapolis, Indiana; Olean, New York; York, Pennsylvania; Charlotte, North Carolina; Parkersburg, West Virginia; Houston, Texas; and Youngstown, Ohio.

The author has served twice on the Employer Verification Panel of Ohio Competency Analysis Profile (OCAP) workshop to establish core competency list for the Ohio Vocational Program on Welding. On May 2, 1992 he conducted a Seminar for Vocational Welding Instructors titled, "Improve Your Skills in Teaching Welding," at the Tolles Technical Center. The seminar was sponsored by the Division of Vocational and Career Education, Ohio Department of Education.

The author was selected by the Adult Director of Central Ohio Joint Vocational School District (Tolles) as an adult education representative to serve on the Tolles Strategic Planning Committee conducted by Taylor and Taylor Educational Consultants in the Spring of 1995.

He retired August 1, 1996 after thirty-one years with Columbia. His love for the art of welding continues with his consulting to companies and schools who want to improve their welding operations.

CONTENTS

CHAPTER 1

MANUAL WELDING INSTRUCTIONS

LEARNING OBJECTIVES

After mastering this chapter, you will be able to:

- Understand the theory of welding
- Explain the requirements for welder certification
- Understand the importance of safety rules
- Describe weld joint types and welding positions

This book has two objectives:
(1) to explain the theory (how and why) of manual welding
(2) to describe some good techniques that welders use that are not
   recorded in the welding literature

If you read and understand each chapter and apply the techniques in the practice exercises, you will be able to pass most companies' performance qualification tests.

Another benefit of mastering these techniques and understanding why they work is that it could open the doors of opportunity to become welding technicians, welding instructors, supervisors, and welding shop owners.

Techniques described in this book have been used to make millions of quality welds. For example, pipeline welds made using these techniques have provided reliable service since the 1930's.

WELDING DEFINED

Before we begin, what is manual welding? Manual welding is the joining of two metallic parts into one by a welder trained and certified in the techniques required to overcome the effects of gravity using the puddle method. Throughout this book, definitions and explanations will be in terms of the craft person's language, which I hope will be more easily understood than the formal terminology of the American Welding Society.

The worth of any prospective welder is their knowledge and ability to make quality welds. Therefore, the object of this book is to teach you how to make manual arc welds that meet most industry standards. By learning these skills, you can immediately start earning a good salary. This is a WIN, WIN, WIN, project. You win, your company wins, and the country wins.

REQUIREMENTS FOR WELDER CERTIFICATION

Before choosing welding as a career, you should understand some of the environmental factors involved in welding. The following is a list of some of the conditions you may have to accept:

- Working in a dirty environment
- Wearing heavy clothing in hot weather
- Doing hard physical work

Good welders tend to have the following attributes:

- Hand-eye coordination
- Self motivation
- Determination
- Quick thinking (observing and responding quickly)
- A habit of following instructions

It is not necessary to have all of the above requirements because determination can make up for other things such as good hand-eye coordination or a good instructor.

If you can accept the above conditions and you are determined to succeed, you will be able to pass most industry weld performance tests by using the techniques described in this book.

WELDER TRAINING TIPS

Instead of assigning one student to each booth, assign two or three. This method not only reduces the amount of capital investment in equipment and facilities, but it accelerates the learning process. As the instructor visits each booth, he can make suggestions to all three students as they observe the puddle. When the instructor is not present, the first two students will observe and critique the third while he or she practices the welding techniques. This method increases safety by providing 100% fire watch.

WELDING JOINTS

Most of the practice exercises will help you learn to make groove welds. If you can make groove welds and understand the theory and techniques, you will be able to make fillet welds. The American Welding Society (AWS) definitions for welding joint types are shown in Figure 1-1. Be sure to learn this terminology before proceeding to the following chapters.

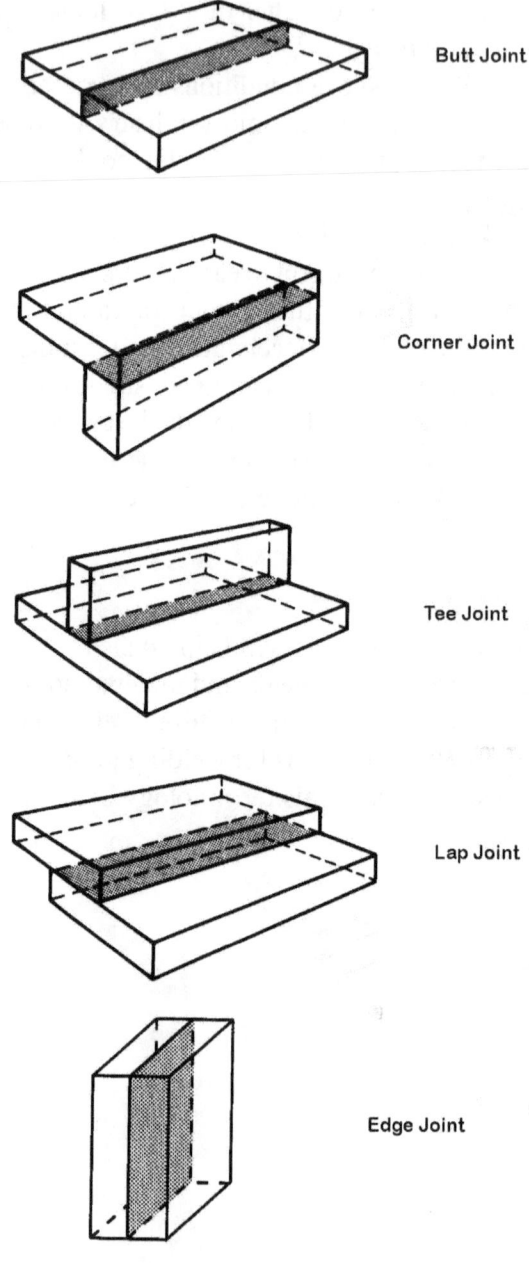

Butt Joint

Corner Joint

Tee Joint

Lap Joint

Edge Joint

Figure 1-1 Joint Types

WELDING POSITIONS

There are different welding positions that you will have to master as you develop your welding capabilities. The progression will be from the easiest position (1G) to the hardest (6G). The American Society of Mechanical Engineers (ASME) and the American Welding Society (AWS) define the welding positions as shown in Figures 1-2, 1-3 and 1-4.

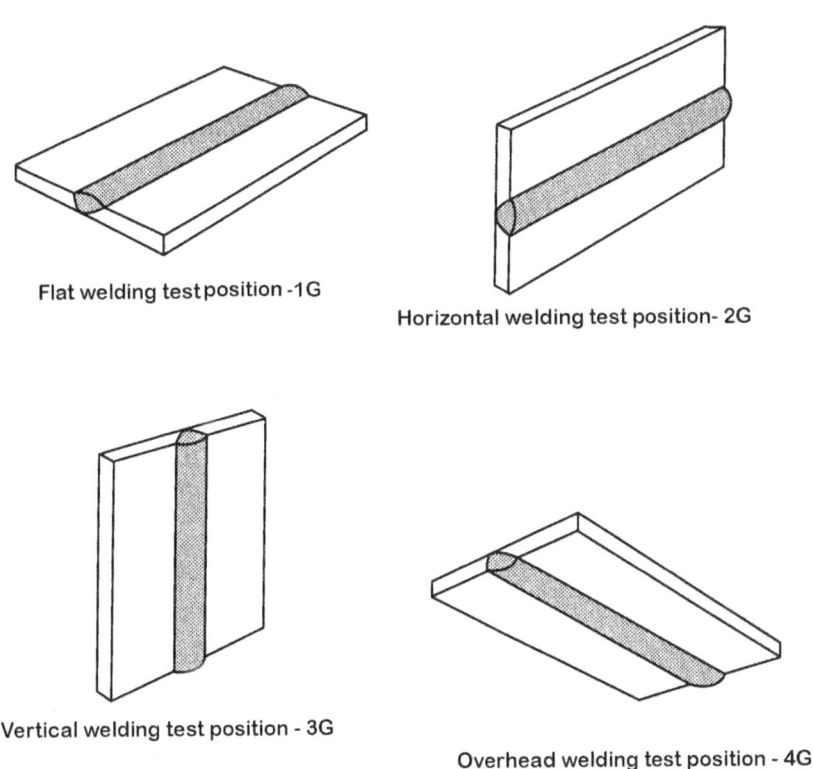

Flat welding test position -1G

Horizontal welding test position- 2G

Vertical welding test position - 3G

Overhead welding test position - 4G

Figure 1-2 Test positions for groove welds in plate

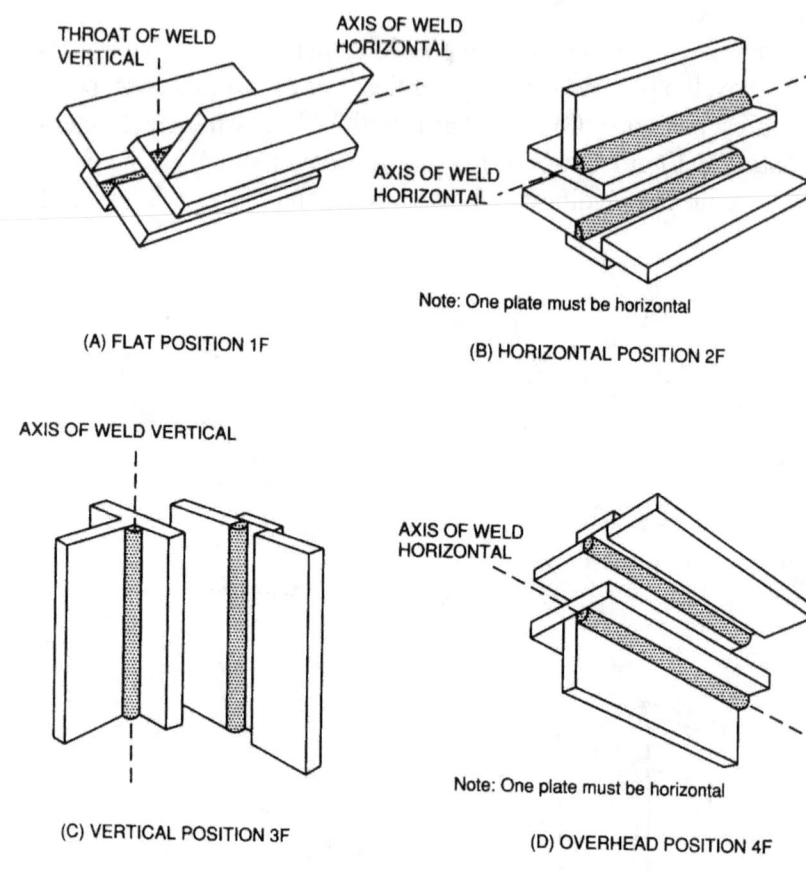

THROAT OF WELD VERTICAL

AXIS OF WELD HORIZONTAL

AXIS OF WELD HORIZONTAL

Note: One plate must be horizontal

(A) FLAT POSITION 1F

(B) HORIZONTAL POSITION 2F

AXIS OF WELD VERTICAL

AXIS OF WELD HORIZONTAL

Note: One plate must be horizontal

(C) VERTICAL POSITION 3F

(D) OVERHEAD POSITION 4F

Figure 1-3 Test positions for fillet welds in plate

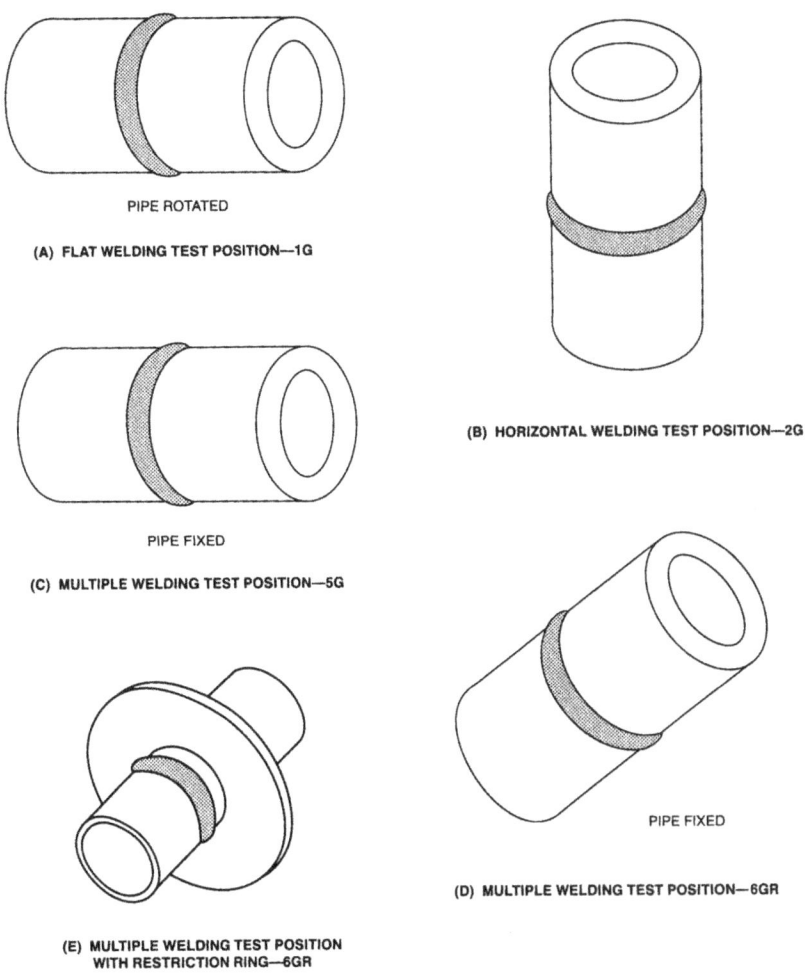

Figure 1-4 Test positions for groove welds in pipe

WELDING PROCEDURES

Welding procedures are defined as the instructions given to the manual welder describing how to make a quality weld joint. The procedure has been tested (qualified) to demonstrate that the given parameters produce quality welds. However, the welding procedures do not specify the techniques. Welding can be compared to cooking. The recipe tells how to make an excellent cake; the welding procedure tells you how to make an excellent weld. However, most welding procedures list fewer details and information than cookbooks because most welding books are written by non welders who are not aware of these techniques.

SAFETY

Manual welding is a safe occupation if you follow the safety rules in publications such as:

- Employer defined safety rules and regulations
- National safety standards such as AWS Z49.1, Safety in Welding and Cutting
- Occupational Safety and Health Act (OSHA) Part 1926 and Part 1912
- Manufacturer's safety rules
- Warning labels

Some chapters list safety rules that will reinforce the importance of practicing good safety and housekeeping rules. A good rule to follow is not to operate any welding equipment without first reading and understanding the manufacturer's safety and operating instructions.

CHAPTER 1 REVIEW QUESTIONS

1. Is welding theory more important than welding techniques to the manual welder?
2. Name five attributes that will help you in learning to make quality welds.
3. Before operating welding equipment, you should read and understand the _____
4. safety and operating instructions.
5. List the American Welding Society welding positions in order of increasing difficulty for groove plate welds.
6. Define the term welding.

CHAPTER 2

ARC WELDING THEORY

LEARNING OBJECTIVES

After mastering this chapter, you will be able to:

- Trace the historical development of the shielded metal arc welding (SMAW) process
- Explain the theory of SMAW process
- Understand the importance of joint design, alignment, and cleaning
- Explain how weld quality is monitored by observation of molten weld pool

INTRODUCTION

Sir Humphrey Davy of England first created an arc between two terminals of an electrical circuit while experimenting with electricity in 1801. Davy exhibited this miniature bolt of lightning at the Royal Institute of England in 1808 where it aroused a good deal of curiosity. The arc was not used for welding until Benardos and Olszewski were issued a British Patent in 1885.The patent was for a process wherein fusion was obtained from an arc between a carbon electrode and the workpiece. In 1890, Russian Nicholas Slavianoff applied for a patent titled, "Electrical Casing of Metals". Slavianoff used a bare metal wire in place of the carbon electrode. This was the beginning of the metal arc welding process.

The shielded metal arc welding process is a process in which the heat for joining is obtained from an electric arc between a covered metal electrode and the workpiece. The groove is filled by the melting of the electrode core. The molten weld pool is protected from the atmosphere by the gases formed from the decomposition of the electrode covering. The molten weld metal is cleaned by the scavenging action of the molten flux. Figure 2-1 shows the essential components of the shielded metal arc welding process. Some of the more common names for SMAW are "stick" welding and "arc" welding.

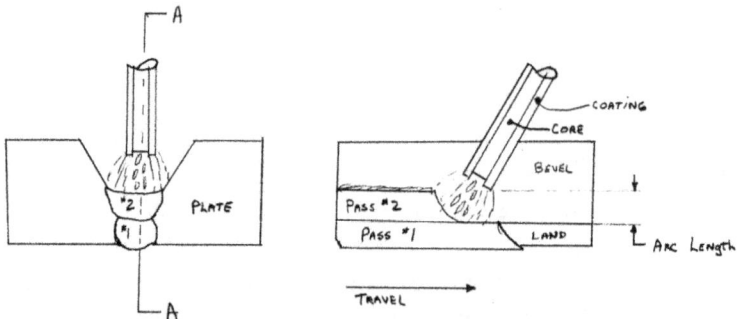

Figure 2-1 Components of Arc Welding

HISTORICAL DEVELOPMENT

In the late 1880's, metal arc welding was discovered independently by N.G. Slavianoff, a Russian, and Charles A. Coffin of Detroit (Coffin later became president of General Electric Company). The carbon electrode used in earlier methods was replaced with a bare metal electrode.

The first arc welder used in the U.S.A. was a German machine imported in 1907. It was a 200 amp, 60 volt generator. In 1917, there were only four established arc welding equipment manufacturers in the U.S. One was the Lincoln Electric Company which was formed in 1895 by John Lincoln to manufacture electric motors. Lincoln made its first variable voltage DC welder in 1907 but was not active in the welding field until after 1912 when James F. Lincoln joined his older brother in the firm. The General Electric Company also entered the welding generator field in 1912 with C. A. Coffin as president.

The Consumers Power Company in Pontiac, Michigan, started using the electric arc method of welding gas transmission and distribution pipelines in 1923. The first job was extending 6" and 8" high pressure lines from a generating plant. The welding current was carried over an obsolete trolley line. In 1925, 50 miles of pipe was installed using DC current generated by several G.E.W.D. - 12 generators mounted on Fordson tractors and belt driven as shown in Figure 2-2. The welding was completed using a 5/32" uncoated low carbon electrode. All welds were roll welded with the welding from 3 o'clock to 12 o'clock position (uphill). Then the pipe was turned 90 degrees resulting in all welding being done on the upper quadrant. The pipe was beveled to 45 degree with zero root face and 1/16" root opening.

Figure 2-2 G. E. Generator

In the 1920's and 1930's, there were many pioneering and innovative individuals who risked everything on an idea. One of these pioneers was Harold (Hal) C. Price. At the age of 33, Price was out of work due to the closure of the Bartlesville Zinc Company in Bartlesville, Oklahoma. While many scoffed at the future of electric welding, Hal Price saw great promise. In 1921, Price borrowed $2,500 and opened the Electric Welding Company in a little shop in Bartlesville, Oklahoma. He started using bare electrodes to repair the bottom of 55,000 barrel tanks for Empire Pipeline Company. The job was so successful that Price closed up the shop in Bartlesville and started contracting tank welding in the field. By 1926, H. C. Price Company was the largest electric welding contractor in Mid-America. In 1928, the Texas Pipeline Company planned the first major electric welded pipeline. The 8 inch oil line was installed from Corsicana to San Augustine, Texas.

Price got the contract for the welding and two foremen and 19 welders went to work. The 18 welding machines were mounted on wooden farm wagons drawn by horses as shown in Figure 2-3.

Figure 2-3 1931 Lincoln Gasoline Generator

Hal Price worked with James Lincoln of Lincoln Electric to further the art of electric arc welding. Hal Price was the first contractor to use Lincoln's new coated welding rod called Shielded Arc, to weld a Cities Service Pipeline Company job in 1930 that was 32 miles long and 20 inches in diameter. In 1931, Price bought 40 new 400 ampere gasoline driven welding machines to burn the 5/16 inch and 3/8 inch heavy-coated welding rods. All of the welds were roll welds against back up rings. In 1933, one of Price's welders suggested eliminating the backup ring completely. Price welded the 8" job without a ring and the 227 mile line was tested for 48 hours with 1,200 pounds of oil pressure and no leaks appeared. This new technique of arc welding is still used today. This discovery by a Price welder has done more to keep the cost of pipelines reasonable than all of the modern automatic pipeline welding machines put together.

ELECTRICAL CIRCUIT FOR ARC WELDING

The fundamental electrical circuit for manual arc welding is shown in Figure 2-4. The circuit consists of the following:

- constant current welding machine
- electrode holder
- ground clamp
- welding cables
- covered electrode
- material to be joined

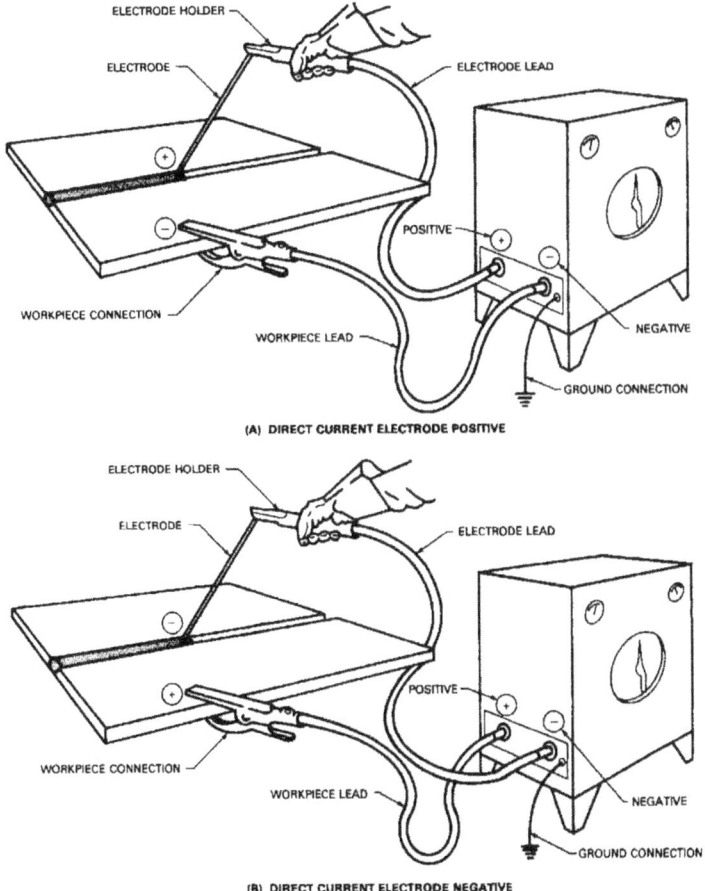

Figure 2-4 Manual Arc Electrical Circuit

The welding operation is started when you place the tip of the electrode quickly onto the work piece and remove it approximately 1/8" above the plate surface. This action starts the current flowing. The distance that the core wire is held above the weld pool (puddle) is called the arc length. The welding operator must feed the electrode toward the liquid puddle at the same rate as the electrode melts (burns). This rate is directly proportional to the previously selected welding current. The welding arc is one of the hottest heat sources, creating a temperature in excess of 9,000 °F (5,000 °C). The intense heat quickly melts the tip of the electrode and the base metal beneath the arc stream. Tiny globules of the molten core wire form on the electrode tip and are transferred through the arc stream into the weld puddle.

To melt the base metal and the filler metal (core wire), a certain amount of current must be provided by the welding machine. As the electrode diameter increases, the current required also increases. The welding current can be either alternating (AC) or direct (DC) depending on the type of electrode used. The current can vary from 80 amperes for 3/32" diameter electrode to 180 amperes for 3/16" diameter electrode.

The importance of the power source and electrode coatings will be discussed separately. The dynamics of the arc welding circuit are affected by the hand and eye coordination; therefore, you must continually measure the arc length with your eyes in order to maintain a constant arc length. Now you know why it requires many hours of practice to become a skilled manual arc welder.

POWER SOURCE

For manual arc welding a constant current type welding machine is used. It can be either alternating current (AC) or direct current (DC) if the corresponding type of electrode is used. Both AC and DC current have features and limitations that must be considered when determining the type of current for a specific application.

Some of the factors that should be considered are the following:

- Arc Starting is easier to start with DC current
- Voltage Drop is lower for AC current over long distances from the power source

- Arc Length - short arc length is easier with DC current
- Arc Blow can be eliminated by changing from DC to AC current
- Welding Position - DC current is better for vertical and overhead positions
- Low Current - low currents operate better when using DC current

In most cases, exception being arc blow or long welding cables, the use of DC current is preferred over AC current. If the only welding machine available is an AC type, then the operator should select an AC electrode type having the operating characteristic desired for the application being welded. If DC current is used, you should check the welding cable connections to assure that the work cable is connected to the negative terminal and the electrode cable is connected to the positive terminal. This is called reverse polarity. Most DC welding electrodes operate on reverse polarity.

JOINT DESIGN AND PREPARATION

The bare wire arc welding process was used during the First World War and rusty fence wire was often the filler metal. The first joint designs were lap welds or square joint design with metal backing strip. The welding was done with large diameter electrodes in the flat (downhand) position. After the development of covered electrodes, the electrode diameters were decreased and operating characteristics were improved. Welding operators experimented with the new electrodes and developed a technique where full penetration could be obtained with the use of backing strips. The square plate edges were beveled to allow access to the root. The first horizontal fixed position welds were made using an uphill travel direction using a 37½ degree bevel angle with *1/16"* root face. The same joint design was used for oxyacetylene forehand welding.

The base material (plate or pipe) was usually beveled with an oxyacetylene torch and the bevel root face formed with a file. In the 1930's, the downhill technique was perfected using a 30 degree weld bevel angle. This resulted in faster travel speeds and decreased welding time which made electric welding competitive with oxyacetylene welding.

ALIGNMENT AND CLEANING

The shrinkage stress from welding will cause the base material to move out of alignment. Therefore, the steel weldments are held in place with various clamping fixtures. These devices can be as simple as a C-clamp or an automatic air-operated holding fixture. The clamping fixture may also hold the root face opening until tack welding is completed.

The cleaning of the base metal next to the weld area, as well as the weld, is of great importance. The General Electric Company issued an "Arc-Welding Manual" in October 1928 that listed the following accessories for cleaning bare electrode arc welds:

- steel-wire scratch brush used for light rust and scale and for cleaning between layers
- chisel, hand or air-operated, used for removing slag from weld deposit, or scale from oxyacetylene cuts
- sand blast may be used for removing rust, scale, grease, and dirt of all kinds
- tongs
- chain block
- wedges
- "c" clamp
- soapstone crayon

ELECTRODE TYPES

The composition of the electrode covering affects the welding arc length, the welding position, the chemical composition of weld metal, the mechanical properties, the bead shape, the surface cleanliness of weld metal, arc stability, internal cleanliness, and operating current requirement.

The materials used in the covering may be listed according to their purpose:

- shielding gas
- arc stabilizers
- slagging ingredients
- deoxidizers

- alloy additions
- binders

Materials used for gas shielding are wood flour, wood pulp, refined cellulose, cotton binders, starch, sugar, and other organic materials. Some of the ingredients used for fluxes and slagging are silica, alumina, clay, iron ore, rutile, limestone, magnesite, asbestos, mica, and other man made minerals such as potassium titanate and titanium dioxide.

Ferro-alloy and pure metals serve as deoxidizers and alloying ingredients while the alkaline earth metals are the best arc stabilizers.

The composition is held together with sodium or potassium silicate binders.

QUALITY IMPROVEMENT OF ARC WELDING

The quality of arc welding has been improved greatly from the early metal arc (bare) welding. The improvements can be classified into two efforts. The manufacturers of welding machines and welding electrodes improved the operating characteristics of the equipment and electrodes. At the same time, the skilled welders were trying different techniques to improve the weld quality. As better techniques were discovered, they were passed on by word of mouth from one generation to the next.

The quality of arc welding was improved by using the visual observation of the weld puddle to control welding variables such as the current and the travel speed. This technique of weld puddle control will be discussed in more detail in the following practice chapters.

Chapter 2 Review Questions

1. What is the difference between shielded metal arc welding and stick welding?
2. _____was the first contractor to use Lincoln Electric's coated electrode to weld Cities Service Pipeline job in 1930.

3. Name the six parts of an arc welding circuit.
4. How can arc blow be eliminated when welding with an AWS E6010 electrode?
5. Is the stick electrode easier to start on AC current?
6. What polarity is used for most DC type of electrodes and is the electrode positive?
7. Why are the materials to be welded held in a clamping fixture?
8. List five of the purposes for the electrode coating.
9. How is the quality of manual welding maintained?
10. Why should the base metal be cleaned before welding and after every pass on a multipass weld?

CHAPTER 3

ARC STRIKING, BEADING, AND WEAVING

LEARNING OBJECTIVES

After mastering this chapter you will be able to:

* Understand the importance of safety rules
* Fabricate holding fixture
* Explain puddle observation and reaction
* Understand electrode movements, work angle and electrode angle
* Make bead on plate welds with and without electrode movement

This chapter is the beginning of a step by step course of arc welding instruction designed to provide a method of learning, whereby, the welding operator is told "how" and "why" these illustrated techniques work. The methods described have been used successfully since the 1930's. However, some of these techniques were passed on by word of mouth from one generation to the next and never recorded in a text book.

This instruction will attempt to share some of these secrets of the trade so that you can learn to make code quality welds in a few weeks of reading and practice.

Practice alone will not make you a good welder. If you practice bad techniques, you will be doing more harm than good because the poor techniques will become comfortable and hard to change. Be patient! Begin your welding journey by mastering the basics one step at a time. Each chapter builds on prior knowledge and the more difficult positions will be easier if you master the basics.

The key ingredient to a successful arc welding career is puddle observation and reaction. I have used this puddle observation technique to instruct adults at the Tolles Technical Center. Adult students from all walks of life learn to weld in 35 hours total instruction time (10 nights for 3½ hours each night). With a minimum of instructor help, you can learn to make quality welds in a relatively short time. You will also know why certain techniques work and be able to solve future welding problems by applying the knowledge you learn. I believe it will be an enjoyable experience for you.

SAFETY FIRST

Before attempting any welding, study the safety precautions listed in Tables 3-1 to 3-3. Welding can be a very enjoyable experience if a few simple safety rules are followed.

Table 3-1 Safety Checklist for Personal Protective Equipment
1   Wear safety glasses with side shields at all times—before welding, during welding, and after welding.
2   Do not wear photo gray glasses because the welding arc will darken the lenses during welding making it impossible to see the puddle
3   Never look at the arc without your helmet down
4   Wear a head shield to protect your eyes and face from the rays of the arc.
5   (larger number) shades for welding at night or in a dark shop area, and use lighter (lower number) for outdoor welding applications. Table 3-2 lists suggested shading for various operations

6   Use clear plastic protective plates on both sides of the dark filter lens.

7   Wear protective chipping goggles when chipping off weld slag. Chip away from your face.

8   All skin should be covered. Wear a long sleeve shirt with the collar buttoned up. Wear pants without cuffs or frayed clothes. Always wear 100% cotton clothing. Do not wear frayed or synthetic fabrics.

9   Wear leather gloves and protective clothing such as an apron, sleeves, etc., to shield against the arc rays and sparks.

Table 3-2 LENS SHADE SELECTOR

Operation	Shade Number
Soldering	2
Torch Brazing	3 or 4
Oxygen Cutting up to 1 inch	3 or 4
Oxygen Cutting 1 to 6 inches	4 or 5
Oxygen Cutting 6 inches or over	5 or 6
Gas Welding	4 or 5
Shielded Metal-Arc Welding	9 or 10

Table 3-3 Safety Checklist for Equipment and Environment

1   Keep work area neat, clean, and dry.

2   Observe normal operating care for electrical hazards.

3   Keep equipment in good, clean, dry condition.

4   Use correct size welding cable—don't overload.

5   Make sure all electrical connections are tight, clean, and dry.

6   Be sure cables, holder, and connections are properly insulated.

7   Cut off power to welder before cleaning machine or making internal adjustment.

8   Do not allow welding cable to wrap around body.

9   Always remove electrode stubs from holders and dispose in a metal can.

10   Never change polarity while machine is under load.

11   Use a non-reflecting welding curtain to protect others in the area from the arc rays.

12   Never strike an arc on a compressed gas cylinder.

13   Don't weld near volatile, flammable liquids, or gases.

14   Remove flammable materials from welding area, or shield them.

15   Don't weld or cut on containers such as drums, barrels, or tanks until you know there is no danger of fire or explosion. (See A.W.S. Bulletin—Safe Practices for Welding and Cutting Containers That Have Held Combustibles.

16   Be sure work area has adequate ventilation—plenty of fresh air. Special precautions are necessary when welding lead, zinc, beryllium copper, or cadmium Always keep head out of arc plume (welding fumes).

Holding Fixture

Since you will be making many practice welds, consideration should be given to the fabrication of a test plate holding fixture. The fixture should be adjustable for all positions and should be simple and easy to use. There are many different designs that use C-clamps and welded nuts with bolts to hold the test plates. Figure 3-1 shows a design that uses a 10" vise grip and pipe.

Figure 3-1 Fabricated fixture for holding practice plate

FUNDAMENTAL OF STRIKING ARC

Fasten the test plate securely in the holding fixture. Adjust the height to a comfortable position and draw lines parallel to one side approximately 1/2" apart. You should be facing perpendicular to the soapstone practice lines. If you are right-handed, you will travel from left to right and if you are left-handed you will travel from right to left. Place a 5/32" diameter AWS 6010 electrode in the perpendicular slot in the electrode holder. Turn on the machine and set the welding current at 150 amps and connect the welding cables for reverse polarity. If you are right-handed, hold the electrode holder in your right hand with the electrode release lever facing upward. Do not grip the electrode holder tight because your muscles will become tense, resulting in cramps and shaking hands. Your left hand may be used to lightly touch your right hand to help steady the electrode.

To start the arc, the electrode tip is quickly brought into contact with the work piece (test plate), then quickly raised until there is a 1/8" gap between the rod and the test plate. One method is to strike the test plate a hard blow with the tip of the rod and allow it to bounce up to establish the 1/8" arc gap. It is very important to strike the arc quickly and not allow the electrode to remain in contact with the work piece. If the electrode remains in contact too long, it will weld itself to the plate. If this happens, quickly release the electrode by squeezing the lever on the electrode holder.

Another method used is to scratch the electrode tip on the plate as you would a match, then lift it quickly to obtain the 1/8" gap. This is usually done slightly in front of where the weld is to begin. Sketches of both methods are shown in Figure 3-2.

Begin by aligning the electrode tip directly above where you want to start the welding. After lowering your hood, the electrode is moved straight toward the plate with one of the above methods.

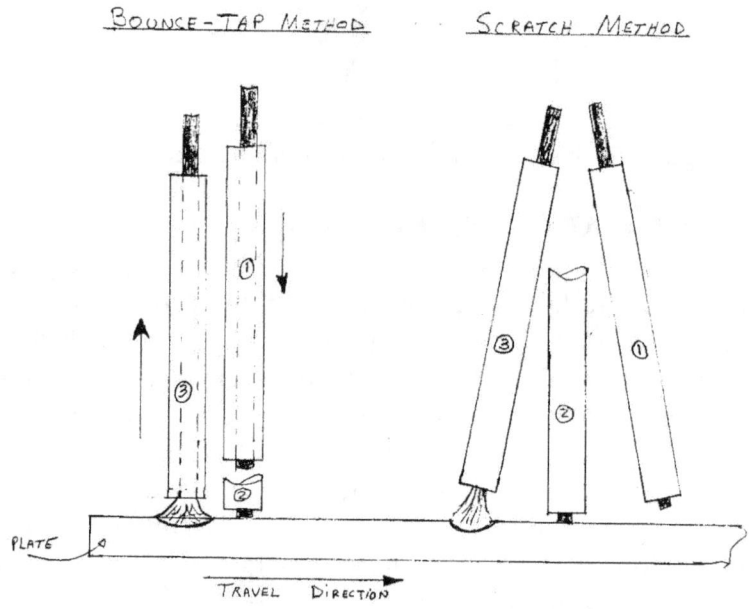

Figure 3-2 Electrode Start Methods

In the beginning, you may feel your way until the tip of the electrode touches the plate. This will usually result in sticking or welding the electrode to the plate which produces a direct short on the welding machine. If this occurs, depress the holder lever with your thumb to release the electrode from the jaws of the holder.

The electrode should be released immediately to avoid unnecessary heating of the welding electrode which will damage the flux coating and also make arc striking more difficult. If the rod will not start, check the ground cables, the electrode connections, and the welding current setting (may be too low). Use a 5/32" diameter electrode because it is easier for beginners to strike an arc than smaller diameter rods.

Practice starting (striking) the arc and maintaining it for a few seconds. The arc length should be held close to 1/8" constant length by feeding (moving) the electrode tip straight toward the weld puddle. If you feed the electrode slower than it is melting (burning), the arc length will increase until it is so long that the arc is extinguished. If you feed the electrode faster than it is burning, the electrode may freeze (weld) to the plate. Your head should be perpendicular to the electrode so that the arc length can be continually observed. If the arc length starts getting longer, the electrode should be fed faster and if the arc length gets too short, the electrode feed should be slowed down. The arc is stopped by rapidly pulling the electrode away from the plate. Continue practicing starts and stops until you are comfortable and confident that the weld can be started and stopped routinely. Figure 3-3 shows the different arc length conditions.

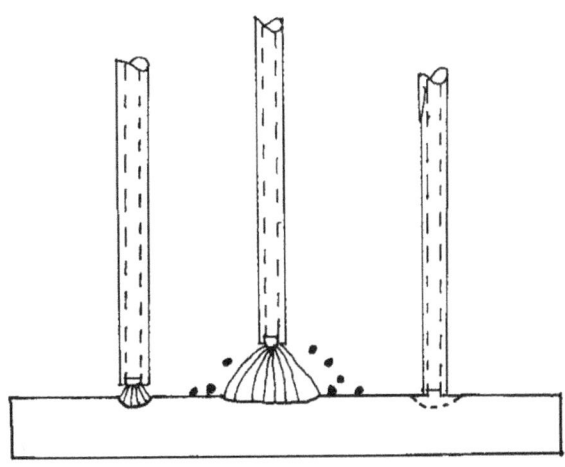

Figure 3-3 Arc Lengths

Puddle Observation and Reaction

The observation of the puddle is the most important technique that you will learn in this book. This technique will decrease the time it takes to learn to arc weld and it will increase the quality of your welding. In general terms, what you will be doing is controlling the following welding variables by observing the weld puddle area:

- welding current
- arc length
- travel speed (slag and fill)
- electrode travel angle
- electrode work angle
- electrode movement (width and direction)

Another amazing benefit of the puddle technique is that it is easier to make good welds than to repair poor welds made using bad techniques.

The welding current is set by observing the liquid puddle. If the puddle is fluid and the surface hopping from the arc force, then you have the correct current setting. Some welders are afraid of high welding currents. Therefore, they make most of their welds too cold. This is bad for two reasons. First, the weld is likely to have areas of lack of fusion, slag inclusion, rough appearance, and incomplete penetration. Secondly, the weld is more difficult to make. I like to use the analogy of painting. If paint is thick, it is difficult to spread and looks terrible. If a paint thinner is added (similar to increasing the weld current) to the paint, it goes on easier, smoother, and bonds better. The same is true for welding. Therefore, use the current settings on the welding machine only as references and set the current by observing the liquid puddle movement (hopping).

The puddle method has the advantage of compensating for current lost in the welding circuit due to long cables, poor connection, and temperature increases in the welding cables. The current should be set on a piece of scrap, before any welding is done on test plates or production parts.

Another advantage of setting the current by observing the puddle is that you do not have to memorize numbers and you can set your current quickly on any new welding machine you may encounter.

Arc length is monitored by glancing at its length every few seconds. Most beginners look at the welding arc from directly over the holder. Never do this because you have no depth perception from this angle. Therefore, you should move your head to the left and down so you are looking perpendicular to the arc length. The length should be slightly less than the electrode core diameter. With practice, the feeding of the electrode will become almost automatic, just like driving a car.

In manual welding, travel speed is difficult to judge unless you use the weld puddle as a gauge. The movement of the electrode forward should be at a slow steady rate. Glance at the back side of the weld puddle to judge the width and height. If the puddle is too low, then the travel speed should be decreased slightly. If it is too full, the speed should be slightly increased. A sudden movement forward will leave a depression (low fill area). If you observe this happening, quickly reverse your travel direction to fill the void, then resume the normal travel speed. A weld is a series of frozen layers and if the travel speed is maintained at an uniform rate, then the finish weld will look uniform. The control of uniform travel speed is a critical element in obtaining a weld free of defects.

ELECTRODE MOVEMENT

The recommended electrode angle and work angle are given. The work angle for these beading and weaving exercises is 90 degree from the plate. The electrode travel angle is 60 degree from the plate with the holder leading the weld puddle. This equates to a drag travel angle of 30 degree.

The electrode movement is zero for the beading practice exercises and about two rod diameters wide for the weaving passes. The weaving should not be too wide and the width should be established by observing the puddle width to obtain uniformity. In groove welds, the bevel width will determine the amount of weaving. A larger weave can be performed in the flat position than in the other fixed positions. The pattern for the weaving motion is shown in Figure 3-4.

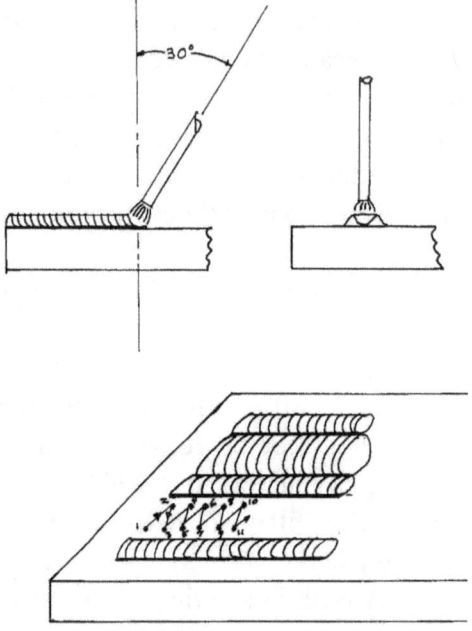

Figure 3-4 Weaving Motion

PRACTICE BEADING

Place 7" x 8" x 1/4" plate into the holding fixture. With soapstone, draw lines parallel to the 8" side about 1/2" apart. Using 5/32" diameter AWS 6010 electrodes make four 8" long beads across the plate. After each pass, the slag shall be removed with a chipping hammer and wire brushed. After making two passes on one plate, a second plate should be marked and two passes welded without weaving using 5/32" diameter AWS 6010 electrodes. Alternate welding between the two plates and if heat is building up, you may have to cool the plates with water.

During all weld beads, the arc length and travel speed should be monitored on a continuous basis. The bead uniformity should improve with each practice bead. The cleaned beads should be inspected for uniformity. After you are comfortable starting and welding bead passes, adjust the welding current up and down to obtain experience with too little or too much current. You may want to change to straight polarity on one bead to determine what it sounds like.

PRACTICE WEAVING

After the two plates (one 6010, one 6011) are beaded, make weave passes using the bead passes as guides. This is done by welding between the bead passes from left to right making one start and stop. Each weld pass should be cleaned and visually inspected for uniformity.

Chapter 3 Review Questions

1. What is the key ingredient to learning to arc weld?
2. What should you read before striking the welding arc?
3. How tight do you grip the electrode holder?
4. Name the two methods used to start (strike) the welding arc.
5. Why should you never look at the welding arc from directly over the electrode holder?
6. List five variables that you can control by observing the weld puddle.
7. Is it easier to weld with a low or high current?
8. How will you control travel speed to obtain a uniform width?
9. What is the travel angle for the beading weld passes?
10. What is the work angle for the beading and weaving passes?

CHAPTER 4

ARC WELDING TECHNIQUES FOR FLAT (1G) AND HORIZONTAL (2G) POSITIONS

LEARNING OBJECTIVES

After mastering this chapter, you will be able to:

- Understand joint design for flat position welding
- Explain the purpose of tack welding and cleaning
- Learn to set welding current and control travel speed with puddle observation
- Control electrode travel angle and work angle for flat and horizontal positions
- Weld V-grooves with and without backup in flat and horizontal positions

Now comes the fun part. You will apply the knowledge gained in Chapter 3 to join two steel plates. Practice welds will be made on a T-joint fabricated from two 1½- inch angle irons. Another exercise will be welding a corner joint between two 1½ by 8 inch long plates. The angle iron may be tack welded in a fixture like the one shown in Figure 4-1.

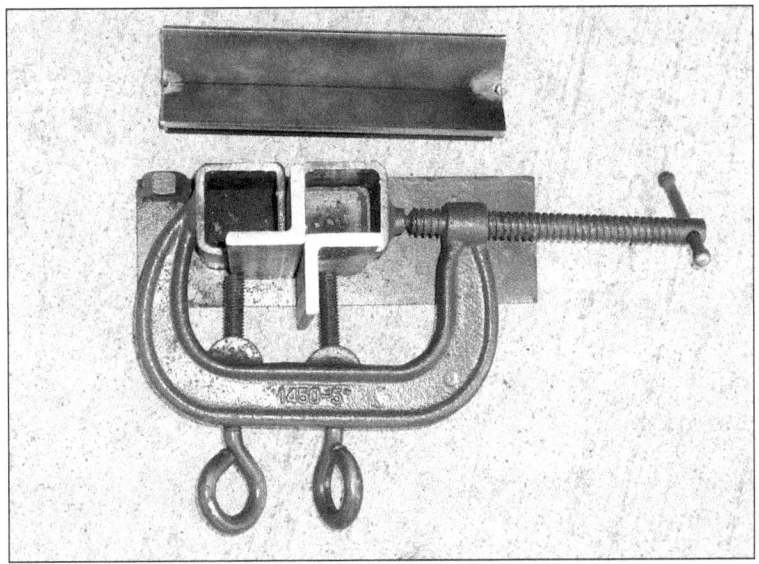

Figure 4-1 T-Joint Tacking Fixture

The T-joint simulates a heavy wall, 45 degree V groove. This allows four practice set-ups with little wasted material. The corner joint simulates a ¼ inch plate thickness with a 45 degree V groove joint. These joints do not require any expensive machining or oxy-acetylene flame cutting equipment.

JOINT DESIGN

The manual arc welding design for butt welds was established many years ago. The two common bevel angles are 30 degree for downhill and 37½ degree for uphill. For the flat and horizontal positions, you will be using a 30 degree bevel. The space between the plates is called the root opening and the nose (blunt end of plate) is called the land. The root and land should be the same dimension.

You will be using a *1/16"* land and root opening or another way of measuring these dimensions is to use a penny. A penny is .062 inches thick. The T-joint emulates a 45 degree bevel with a steel backup. This is easier to make than an open root V groove joint. After mastering the T- joint, the corner joint will be tack welded with a 30 degree bevel, 0" land and *1/16"* space. This will provide experience on welding two plates requiring **full** penetration. Figure 4-2 shows the design for the T-joints to be used in this chapter.

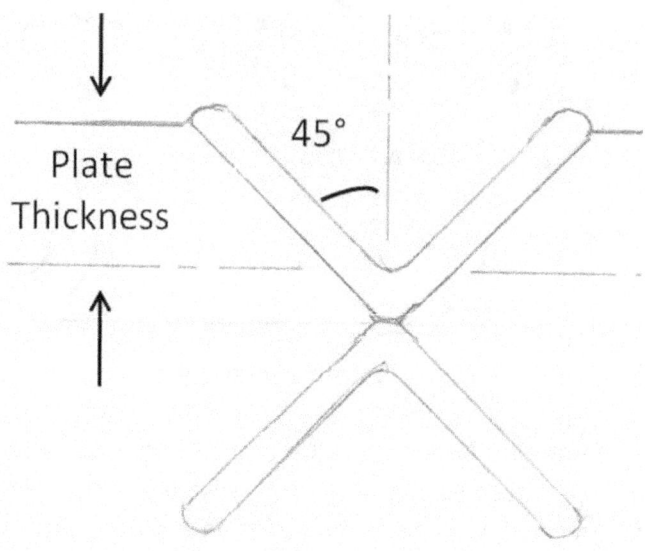

Figure 4-2 T-Joint Design

ALIGNMENT AND TACK WELDING

Alignment is a critical part of preparation for welding. The purpose is to hold the two plates together at an uniform root opening for corner weld joint and to align the T -joint plates with 90 degree included angle. The root opening is gaped with a spacer tool fabricated from an old hacksaw blade with one end ground to form a wedge shape.

Each plate of the corner joint is positioned at a 30 degree angle from the vertical with a *1/16"* root opening. A flatten angle iron can be used as shown in Figure 4-3 to provide a fixture for tack welding the corner joint test plates. If the plates are tack welded with 0 root opening, the root opening can be opened with a hack saw.

Figure 4-3 Home-made fixture
for tack welding the corner joint

CLEANING

The manual arc welding process has some cleaning potential due to the scavengers in the electrode coating but it is always better to clean the bevels and weld area next to the bevels by filing, grinding, or machining before the plates are tack welded.

This initial cleaning is well worth the effort. The bevel surfaces should be smooth and regular with all moisture, rust, scale, paint, oil, or other materials which may be detrimental to the finished weld removed. After each pass, the slag should be removed with a hand wire brush or power wire brush. The excessive bead thickness shall be cut down with a power grinder or round nose chisel before making the next weld pass. You should strive to make each fill pass a uniform width and height.

ELECTRODE TYPE, SIZE, AND ARC LENGTH

The AWS E-6010 electrode is used for welding in all positions. If a direct current welding machine is not available, the AWS E-6011 may be substituted for the E-6010. Most of the operating character-istics are the same for AWS E- 6010 and E-6011. The alternating current used for the E-6011 electrode is harder to start and the sound of the arc is harsh. The arc blow (arc deflection) sometimes encoun-tered with AWS E-6010 is not a problem when using alternating cur-rent. Another advantage for AC welding is the cost of the alternating current machines is less than the direct current type.

Lincoln Electric specifications for operating procedures, mechanical properties and applications for examples of E-6010 and E-6011 electrodes are found at the end of the chapter in Figures 4-7 and 4-8.

A small diameter electrode will be used in this chapter since you should not be having difficulty starting and maintaining the welding arc. All welds will be made using 1/8" diameter elec-trodes. The welding current will be set by observing the movement of the weld puddle. After setting the current with the puddle, have someone record your welding amperage. It should be approximately 30 amperes less than that used on the 5/32" diameter. The 1/8" elec-trode should be placed in the electrode holder perpendicular to the handle or in the 45 degree slot. The 45 degree slot is preferred since it decreases the distance from the electrode holder to the test plate.

While welding, the arc length must be monitored continuously. It should be maintained as close as possible but the electrode should never be dragged on the plate. Always direct the arc toward the root of the joint. Use a long arc when starting then shorten it for welding. Remember to move your head off to the side in order to observe the arc length from the side view (looking perpendicular to arc length). Lincoln operating data for AWS E-7018 is found in Figure 4-9.

WELDING CURRENT

Before welding on any production or test plates, you should set the welding current on a scrap piece of steel. If the arc will not light, check the following: ground connections, machine power on, elec-trode end clean of flux, electrode holder gripping properly, and scrap plate is free of insulating material such as rust, scale, coatings, or concrete.

Since you are using a 1/8" diameter electrode, estimate the current by converting the 1/8" to the decimal equivalent of .125 and throwaway the decimal. Therefore, 125 amperes is a good starting place to set your machine. Another rule of thumb is to add or subtract 30 amperes for each 1/32" increase or decrease from 1/8". For example, 3/32" diameter electrode equals 95 amperes and 5/32" diameter electrode equals 155 amperes.

After the arc is established, shorten the arc length to 1/16" and observe the weld puddle. If the puddle is jumping and fluid, the current is set properly. If the puddle is depressed and unstable (spatter), you may have the current too high. Stop the arc and look at the depression left at the stop. The deeper the depression, the higher the current. If the arc was hard to start and very sluggish (not fluid), you may have to increase the current.

Remember, welding is just like painting, thick paint is difficult to apply and does not bond well. Practice setting the current too high and too low so you can establish reference points for future current settings.

TRAVEL SPEED

The key to making uniform weld passes is to continually observe the liquid puddle. The forward travel is determined by the shape of the liquid puddle.

After establishing the welding arc at the start location, hesitate until the top edge of the weld puddle touches the bevels then start your forward travel. The movement forward should be smooth and gradual. If the travel speed is too slow, the liquid metal will move up the bevels further than at the start. If the travel speed is too fast, the liquid metal will not fill the bevel the same height as the start or may leave a void (low spot) in the weld pass. After a little practice, your weld beads will look smooth. If the weld joint is wider in one area, you will observe this in the liquid puddle. You then need to slow your travel speed until the bevel is filled to the proper height. Automatic welders are set at a fixed travel speed, resulting in low spots where the bevels are wider but a manual welder can slow down and fill in the low areas.

The travel speed for the first pass of the open root corner joint is more difficult than the first pass on the T - joint or the fill and cap passes. The purpose of the first (root) pass is to obtain full penetration without leaving a hole. A key hole is formed at the start which indicates that each bevel is being melted. As the keyhole increases in size, the electrode is whipped approximately three rod diameters forward through an arc to increase the arc length. The electrode is then quickly returned to the keyhole. Some welders call this technique stepping. This sequence is repeated until the root bead is completed. Even though the puddle is a reduced size, it should be observed as well as the molten keyhole. As always, the bead should be cleaned with a chipping hammer and wire brush before the next pass is started.

ELECTRODE MOVEMENT
T-Joint
The first three passes on the T-joint in the flat and horizontal positions are stringer beads. That is, they are made without any lateral movement (oscillation) from bevel to bevel.

The fourth pass in the flat position is oscillated about three rod diameters using a saw tooth motion. The work angle for the flat position is 45 degree for the T -joint. The work angle varies with each pass when welding in the horizontal fixed position.

In the horizontal position, the first pass work angle is 45 degree, the second pass work angle is 60 degree and the third pass work angle is 30 degree. The travel angle for both the flat and horizontal positions is 30 degree drag. That is, the electrode holder is tilted toward the direction of travel about 1/3 distance from the vertical.

Figure 4-4 illustrates the work and travel angles for the flat position and Figure 4-5 illustrates the work and travel angles for the horizontal position.

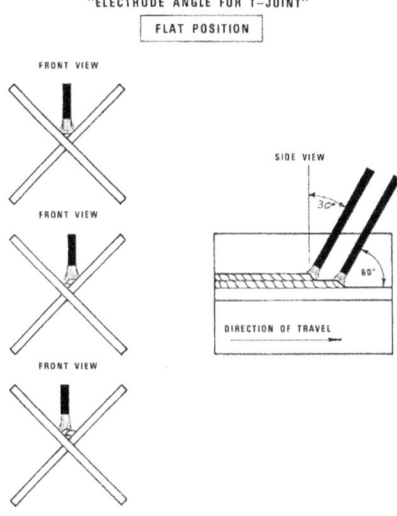

Figure 4-4 Electrode Angles for Flat Position (1G)

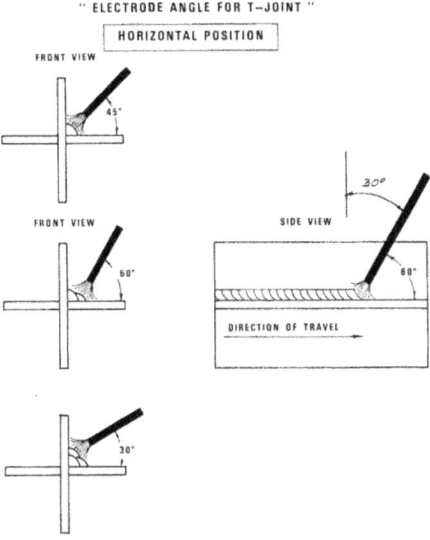

Figure 4-5 Electrode Angles for Horizontal Position (2F)

Corner Joint

The corner joint is made with a 30 degree work angle and a 30 degree drag travel angle. The first (root) pass is made using the stepping technique with a keyhole while the filler passes are made with the same open arc technique used for the T -joint welds. Figure 4-6 shows the travel and work angles.

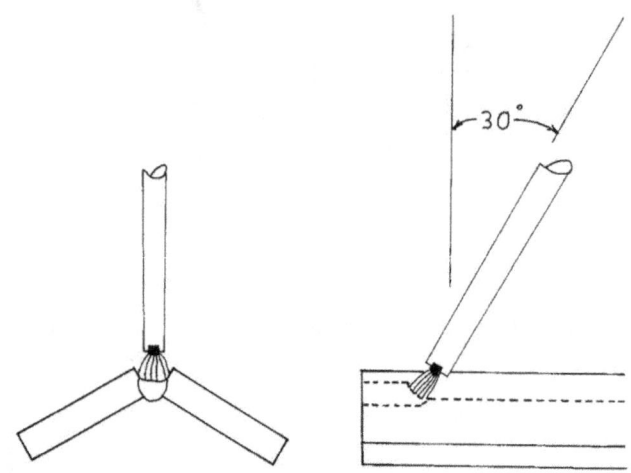

Figure 4-6 Electrode Angle for Flat Position Corner Joint

The electrode angle is important because the electrode core metal is propelled from the electrode tip straight into the puddle. This area is where the electrode arc force is greatest and where the maximum penetration will be achieved. That is why an electrode angle is established to direct the arc energy into the root area. This arc force can, also, be used to support the liquid weld against the universal force of gravity as you will see in fixed position welding. If you understand this theory of gravity, you can determine the electrode angles without referring to a book or your notes.

Welding Technique

The first three passes are made using a stringer bead technique while the fourth pass is made with the weaving technique. The first pass on the comer joint V groove weld is made with an open arc stepping technique.

Visual Inspection

You should check each welding pass after cleaning. If weld discontinuities or irregular bead profiles are made, you should try to remember the circumstance involved in the area where the irregular bead occurred. Table 4-1 is a guide to help you determine the cause of some weld discontinuities found in flat position and horizontal position welds. Remember, your job is not completed until the weld is inspected and accepted.

TABLE 4-1 Discontinuities and Cause	
Discontinuities	**Cause**
Spatter	long arc length, too high current
External undercut	long arc length, incorrect work or travel angle
Low fill	travel too fast
High fill	travel too slow
Misplaced bead	incorrect work or travel angle
Incomplete penetration	root opening too tight - no keyhole
Slag	abrupt changes in travel speeds
Porosity	long arc length, improper cleaning

After some practice using the liquid puddle to control welding variables, you will be producing consistently good quality welds and will enjoy making them.

PRACTICE T-JOINT FLAT POSITION (IF - BACKUP)

Make welds as described. Always clean and inspect each pass. If the T-joint specimen exceeds a temperature of 350 °F, pick up the practice joint with tongs and stir it in a tank of water. The specimen should be dried before the welding is continued.

PRACTICE T-JOINT HORIZONTAL POSITION (2F)

Make welds as described. Always clean each pass and do not overheat. A good practice is to use two specimens so one can be cooling while you weld on the other one.

FLEETWELD® 5P AWS: E6010 Fast Freeze, Out-of-Position Pipe Welding, Mild Steel Stick Electrode

Fleetweld 5P is a great choice for welding on dirty, rusty, greasy or painted steel — especially in vertical or overhead applications.

ADVANTAGE LINCOLN

• All-position, particularly good for vertical and overhead.

• Light slag with little slag interference for easy arc control.

• Deep penetration with maximum admixture.

• Capable of x-ray quality welds, out-of-position.

• Manufactured under a quality system certified to ISO 9001 requirements.

TYPICAL APPLICATIONS

• Tolerates galvanized, plated, dirty, painted or greasy steel which cannot be completely cleaned.

• Pipe welding — cross country and in-plant.

• Joints requiring deep penetration such as square edge butt welds.

• Repair welding.

WELDING POSITIONS

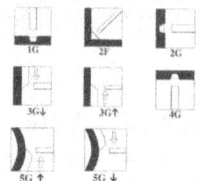

CONFORMANCE

AWS A5.1[n]: E6010
ASME SFA-5.1: E6010
Lloyd's: Grade 3M
ABS: E6010
CSA W48: E4310
[n]See Note 2 on page 44.

MECHANICAL PROPERTIES [n] - As Welded per AWS A5.1-91

	Yield Strength psi (MPa)	Tensile Strength psi (MPa)	Elongation (%)	Charpy V-Notch ft•lbf (Joules) @ -20°F (-29°C)
Required AWS E6010	48,000 (331) min.	60,000 (414) min.	22 min.	20 (27) min.
Test Results As-welded	48,000 - 67,000 (331 - 460)	60,000 - 76,000 (414 - 524)	22 - 33	20 - 71 (27 - 96)
Stress-relieved[n] 1 hour @ 1150°F (620°C)	48,000 - 61,000 (331 - 420)	62,000 - 69,000 (427 - 475)	28 - 36	50 - 55 (68 - 76)

[n] Typical all weld metal.   [n] Data provided for information only – not part of AWS classification.

DIAMETERS / PACKAGING

Diameter Inches (mm)	5 Lb. (2.3 kg) Carton (40 Lb. Master)	10 Lb. (4.5 kg) Easy Open Cans (60 Lb. Master)	50 Lb. (22.7 kg) Easy Open Cans
3/32 (2.4)	ED021212	ED010210	ED010211
1/8 (3.2)	ED021213	ED010202	ED010203
5/32 (4.0)		ED010215	ED010216
3/16 (4.8)			ED010207
7/32 (5.6)			ED010219
1/4 (6.4)			ED010200

DEPOSIT COMPOSITION [n] - As Required per AWS A5.1-91

	%C	%Mn	%Si	%S	%P
Requirements AWS E6010	Not Specified				
Test Results	.08-.15	.35-.55	.15-.25	.010-.020	.005-.010

[n] Typical all weld metal.

TYPICAL OPERATING PROCEDURES

Polarity	Current (Amps)					
	3/32" (2.4mm)	1/8" (3.2mm)	5/32" (4.0mm)	3/16" (4.8mm)	7/32" (5.6mm)	1/4" (6.4mm)
DC+	40-70	75-130	90-175	140-225	200-275	220-325

LINCOLN ELECTRIC
THE WELDING EXPERTS

Stick

www.lincolnelectric.com

Figure 4-7 Lincoln Electric specifications for AWS E-6010

FLEETWELD® 180 AWS: E6011 Fast Freeze, Out-of-Position Mild Steel Stick Electrode

Got a small AC welder? Here's your electrode! Fleetweld 180 offers excellent arc stability for excellent performance with power sources as low as 50V open-circuit voltage (OCV). A great stick electrode with the ability to start easily on low open circuit voltage welders.

ADVANTAGE LINCOLN

- An all-position electrode particularly good for vertical and overhead.

- Light slag with little slag interference for easy arc control.

- Deep penetration with maximum admixture.

- Manufactured under a quality system certified to ISO 9001 requirements.

TYPICAL APPLICATIONS

- Great for use with small AC welders.

- Excellent for sheet metal welding on edge, corner and butt joints.

- Plated, dirty, painted or greasy steel which cannot be completely cleaned.

- All-position welding.

WELDING POSITIONS

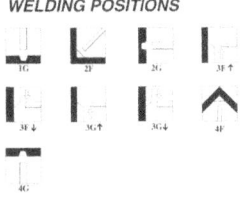

CONFORMANCE
AWS A5.1: E6011
ASME SFA-5.1: E6011
CSA W48: E4311

MECHANICAL PROPERTIES [1] - As Welded per AWS A5.1-91

	Yield Strength psi (MPa)	Tensile Strength psi (MPa)	Elongation (%)	Charpy V-Notch ft•lbf (Joules) @ -20°F (-29°C)
Required AWS E6011	48,000 (331) min.	60,000 (414) min.	22 min.	20 (27) min.
Test Results As welded	48,000 - 70,000 (331 - 480)	60,000 - 84,000 (414 - 579)	22 - 35	20 - 53 (27 - 72)

[1] Typical all weld metal.

DIAMETERS / PACKAGING

Diameter Inches (mm)	5 Lb. (2.3 kg) Carton (40 Lb. Master)	10 Lb. (4.5 kg) Easy Open Cans (60 Lb. Master)	50 Lb. (22.7kg) Easy Open Cans
3/32 (2.4)	ED021224	ED010108	ED010110
1/8 (3.2)	ED021225	ED010103	ED010105
5/32 (4.0)	ED021226		ED010114

TYPICAL OPERATING PROCEDURES

Polarity	(Current) Amps 3/32" (2.4mm)	1/8" (3.2mm)	5/32" (4.0mm)
AC	40-90	60-120	115-150
DC+	40-80	55-110	105-135
DC-	40-80	55-110	105-135

NOTE: Preferred polarity is listed first.

DEPOSIT COMPOSITION [1]

	%C	%Mn	%Si	%S	%P
Requirements AWS E6011		Not Specified			
Test Results	.10-.18	.40-.70	.25-.50	.005-.020	.005-.015

[1] Typical all weld metal.

Stick
www.lincolnelectric.com

Figure 4-8 Lincoln Electric specifications for AWS E-6011

EXCALIBUR® 7018-1 MR AWS: E7018-1 H4R Low Hydrogen, Mild Steel Stick Electrode

When the job involves critical, out-of-position welding, reach for Lincoln Electric's Excalibur 7018-1 MR. It offers a beautifully clean weld puddle, uniform slag follow, and superior wash-in with no undercutting. Also great for welding on steels with marginal weldability.

ADVANTAGE LINCOLN

- Designed for welding mild steel, low alloy steels and steels of poor weldability.

- Capable of x-ray quality welds.

- Ability to tie-in to side walls without undercutting, especially for critical out-of-position applications.

- Clean weld puddle and uniform slag follow make it easy for the welder to "see" and carry the puddle.

- Manufactured under a quality system certified to ISO 9001 requirements.

TYPICAL APPLICATIONS

- Structural steels and bridges.

- All position welding of mild steels, and some high strength, low alloy steels.

- Tolerates steels with poor weldability, such as high sulfur and high silicon steels.

- Welding of piping, fittings, and tie-ins in the petrochemical and power generation industries.

- Applications requiring -50°F (-45°C) toughness properties.

WELDING POSITIONS

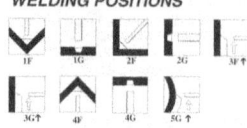

CONFORMANCE

AWS A5.1: E7018-1, E7018-1 H4R
ASME SFA-5.1: E7018-1, E7018-1 H4R
ABS: E7018M, 3, 3YH5
Lloyd's: 3M, 3YMH5
DNV: 3YH5
GL: 3YH5
BV: 3YHHH
CSA W48: E4918-1

MECHANICAL PROPERTIES - As Welded per AWS A5.1-91

	Yield Strength psi (MPa)	Tensile Strength psi (MPa)	Elongation (%)	Charpy V-Notch ft•lbf (Joules) @ -50°F (-46°C)
Required AWS E7018-1 H4R	56,000 (399) min.	70,000 (482) min.	22 min.	20 (27) min.
Test Results As-welded	58,000 - 79,000 (399 - 545)	70,000 - 91,000 (482 - 627)	22 - 35	20 - 130 (27 - 176)
Stress relieved[1] 1 hour @ 1150°F (620°C)	56,000 - 72,000 (386 - 496)	73,000 - 86,000 (503 - 593)	29 - 36	125 - 263 (169 - 356)

[1] Data provided for information only – not part of AWS classification.

DIAMETERS / PACKAGING

Diameter Inches (mm)	10 Lb. (4.5 kg) Easy Open Cans (60 Lb. Master)	50 Lb. (22.7 kg) Easy Open Cans
3/32 (2.4)	ED028701	ED028700
1/8 (3.2)	ED028703	ED028702
5/32 (4.0)	ED028705	ED028704
3/16 (4.8)		ED028706
7/32 (5.6)		ED028919
1/4 (6.4)		ED028920

TYPICAL OPERATING PROCEDURES

Polarity	3/32" (2.4mm)	1/8" (3.2mm)	5/32" (4.0mm)	3/16" (4.8mm)	7/32" (5.6mm)	1/4" (6.4mm)
DC+	70-110	90-160	130-210	180-300	250-330	300-400
AC	80-120	100-180	140-210	200-300	270-370	325-420

Current (Amps)

NOTE: Preferred polarity is listed first.

DEPOSIT COMPOSITION - As Required per AWS A5.1-91

	%C	%Mn	%Si	%S	%P	%Cr	%Mo	%Ni	%V
Requirements AWS E7018-1 H4R	Not Specified	1.60 max.	.75 max.	Not Specified	Not Specified	.20 max.	.30 max.	.30 max.	.08 max.
Test Results	.04-.08	.80-1.50	.20-.65	.005-.015	.010-.020	.01-.06	.08-.25	.01-.05	.001-.010

DIFFUSIBLE HYDROGEN - As Required per AWS A5.1-91

	(ml/100g weld deposit)
Requirements AWS E7018-1 H4R	<4 ml
Test Results 3/32" (2.4 mm)	2.1
1/4" (6.4 mm)	3.3

LINCOLN ELECTRIC
THE WELDING EXPERTS

Stick

www.lincolnelectric.com

Figure 4-9 Lincoln Electric specifications for AWS E-7018 H4R

Chapter 4 Review Questions

1. What type of weld joint does a T-joint make?
2. If the root opening is 1/16 inch, what is the land dimension?
3. A _____ is used to increase the root opening.
4. List five substances that should not be left on the weld bevels.
5. What is a good starting current for a 1/8 inch electrode diameter?
6. How much current should you add for a 1/32 inch increase in electrode diameter?
7. How do you determine if full penetration is obtained when making an open root corner joint?
8. What is the electrode work angle for the first pass of a horizontal T-joint weld?
9. What is the electrode travel angle for a horizontal fillet weld?
10. The first pass of the open root corner joint is made using the _____ technique.

CHAPTER 5

ARC WELDING TECHNIQUES FOR THE VERTICAL DOWN (3G) POSITION

Learning Objectives

After mastering this chapter, you will be able to:

- Understand joint design for vertical down using E-6010 or E-6011
- Explain purpose of tack welding and cleaning
- Learn to set welding current and control travel speed by observing weld puddle
- Control electrode travel angle and work angle when welding vertical down
- Weld full penetration V-grooves with and without backup in the vertical down position

This chapter is your first attempt at supporting a liquid weld puddle against the pull of gravity with an electric arc. This will be accomplished by following the recommendations given in the following pages. You will find that the electrode angles and arc length are more critical than in the flat position. The weak liquid puddle must be supported with the force of the welding arc to balance the force of gravity.

The corner joint will be used to learn the drag technique to make the first pass vertical down. This drag method is used to weld oil and gas pipelines but you will find it a viable method for welding open root, groove joints on plate.

JOINT DESIGN

The joint design will be a 30 degree bevel with 1/16" root opening. As mentioned in Chapter 4, the root opening can be measured with a penny or a hacksaw blade. Since all root and hot passes will be made using the downhill method, all of the joint designs have a 30 degree bevel To practice welding plates with over 1 ½ " thickness, the T -joint will be used which simulates a 45 degree bevel angle. The open root welds will be made using the corner joint test plates with 30 degree bevels. If an oxyacetylene beveling machine is available, the open root welds can be made on 30 degree beveled plate instead of the corner joint plates. If a beveled plate is used, a 1/16" land must be filed to blunt the 30 degree bevel.

ALIGNMENT AND TACK WELDING

The T -joint and corner joint practice plates should be aligned and tack welded as shown in Chapter 4 under alignment. Care should be taken to make sure that the joint design is correct; that is, 1/16" root opening and 1/16" land for beveled plate. If the plates are not aligned properly then it is difficult for you to determine your progress. Skilled welders will have difficulty making quality welds if the joint fit-up design is not uniform.

CLEANING

The bevel surfaces should be smooth and regular with all contaminates such as moisture, rust, scale, primer, and oil removed. The moisture can be removed with heat from an oxyfuel torch and the rust and scale can be removed with a grinder, sander, or file. All weld passes should be cleaned to remove the weld slag. This can be done with a hand or a power wire brush. A chipping hammer may be helpful if a power wire brush is not available. The plate bevels should be cleaned before any welding is done.

ELECTRODE TYPE, SIZE, AND ARC LENGTH

Electrode Type

Your welding skills will be expanded by learning to use the AWS E-6010 or AWS E-6011 electrodes and the downhill technique, The E-6010 electrode will be used with reverse polarity on all passes except the drag root pass which may be changed to straight polarity if the drag technique is too difficult to learn using reverse polarity. Another advantage of straight polarity for the root pass is on thin wall plate (less than *3/16"* thick). Never forget to switch the polarity to reverse polarity for all other passes. The straight polarity cannot be used with an open arc technique.

Electrode Size

A 1/8" diameter will be used on the .250 inch thick plate. For different wall thickness other electrode diameters may be used. For example, for thinner plates, a *3/32"* diameter may be used and on thicker plate a *5/32"* electrode may be used, Refer to Table 5-1 for a list of plate thickness and electrode sizes with approximate heat settings.

TABLE 5-1 T-Joint Heat Settings for Vertical Down (3G) Position				
Electrode Diameter	Plate Fraction	Thickness Decimal	Amps	Puddle Surface
3/32	1/8	.125	90	Jumping
3/32	1/8	.125	60	Smooth
1/8	3/16	.188	120	Jumping
1/8	3/16	.188	90	Smooth
5/32	1/4	.250	160	Jumping
5/32	1/4	.250	120	Smooth
3/16	3/8	.375	180	Jumping
3/16	3/8	.375	130	Smooth

Arc Length

As in all arc welding, the arc length shall be monitored at all times and maintained at a fixed length. The arc length is automatically controlled when using the drag technique since the electrode coating will be in contact with the weld bevel. The arc is nearly invisible since it is under the electrode core. You will note that most of the electric arc light is inside the weld joint. Immediately after starting the welding arc, the electrode is pushed against the bevel and slowly dragged forward. The electrode travel angle is 0 degree and the work angle is 90 degree. If the keyhole under the electrode becomes too large, the travel angle should be changed to a 10 to 20 degree drag angle to decrease penetration and decrease the keyhole size. The travel angle can be changed to a 5 degree push angle to increase keyhole size and increase penetration.

In the 3F position, the arc length for the second pass or hot pass and remaining filler passes is longer than that used in the flat position. This increase in arc length will increase the arc force to balance the liquid weld puddle against the pull of gravity. The longer arc length is maintained at the same length (1/8") for all of the filler passes.

WELDING CURRENT

As mentioned previously, you should set the welding current on a piece of scrap before tacking the practice plates. A good practice is to keep a note pad of the settings you have used successfully in the past. If this is your first vertical down weld, you can start with the welding heat you used for flat position welds. On the scrap plate, strike the arc and observe the puddle. If it is fluid, you have a good heat to start welding. You can set the current just like the open arc flat position welds, then drag the electrode across the scrap. If the electrode snuffs out, raise the heat and if it burns a hole, reduce the heat. The heat can be controlled by the electrode travel angle for the drag technique as discussed in the section on arc length.

For the filler passes, the current should be raised about 10 amps after the drag root pass. That is why pipeliners call the second pass the hot pass. When setting the open arc filler pass current, the weld crater at the end of each pass should be depressed approximately

1/16". A typical 1/8 inch diameter E-6010 electrode may be set at 115 amperes and the hot pass at 125 amperes. Review Table 5-2 for typical welding heats for various electrode diameters using the downhill technique.

TABLE 5-2 Corner Joint Heat Setting

Position	Electrode Diameter	Plate Fraction	Thickness Decimal	1st Pass Amps	2nd Pass Amps	Fillers	Cap
	3/32	1/8	. 125	80	90	90	70
Vertical Down	1/8	3/16	.188	125	130	130	100
	5/32	1/4	.250	150	160	160	130
	3/16	5/16	.312	160	170	170	140

TRAVEL SPEED

The travel speed for the drag technique is controlled differently than open arc weld passes. This is because the arc is buried and the internal keyhole melting is not visible. The electrode is pushed against the weld bevel and slid forward (downhill) as it is burned. The drag technique is more of a feel technique than a hand-eye coordination. The keyhole may or may not be visible. If it is visible, it may be viewed at the back side of the electrode that is opposite the direction of travel. The disadvantage of seeing the keyhole is that if it increases in size it will blow a hole in the root bead. The disadvantage of a buried arc (not visible) is that if it gets too small, the arc may be extinguished. This requires grinding and an extra start and stop. Therefore, the technique you use is a personal choice. Both techniques work.

The second pass (hot pass) is a high amperage pass to remove the parallel slag lines that result from dragging the root pass. These slag lines are why you only drag the electrode in an open root joint design. The travel speed for the second pass (hot pass) is determined by watching the weld puddle. The current should be high enough to depress the puddle and the completed pass should blend smoothly into the bevels and be concave in shape.

The remaining fill passes are controlled by shaping the puddle the same as the hot pass. The fill passes will be wider but the puddle shape is maintained by the electrode force and weaving frequency which keeps the liquid metal from flowing around or past the electrode.

ELECTRODE MOVEMENT

The electrode movement for the ¼" thick corner joint weld is zero for the first pass since the electrode coating is in physical contact with the weld bevels. The second pass usually will be made with no oscillation or just a slight sideway movement. The third pass will be oscillated more than the second pass. Figure 5-1 shows the oscillation or weaving patterns for each pass. There is a limit on the maximum oscillation width when traveling downhill. This is because the liquid puddle may flow around the electrode due to the force of gravity overcoming the support of the arc force.

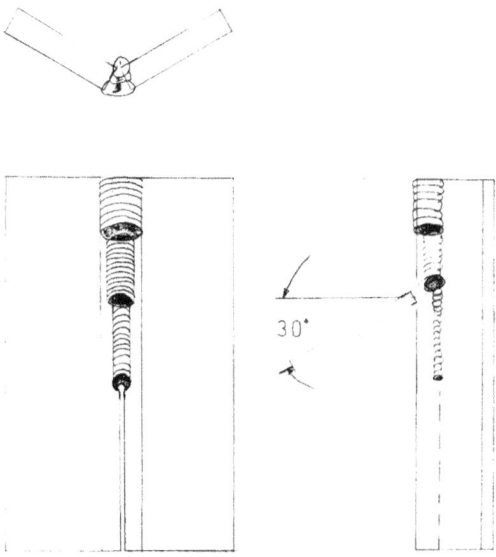

Figure 5-1 Electrode Travel Angle for Corner Joint

The T-joint is made with three passes using the open arc down-hill method. The electrode movement for the filler passes will be greater than the corner joint due to the increased bevel angle of 45 degree.

The travel angle is 30 degree drag and the work angle is 45 degree from the plate surface shown in Figure 5-2.

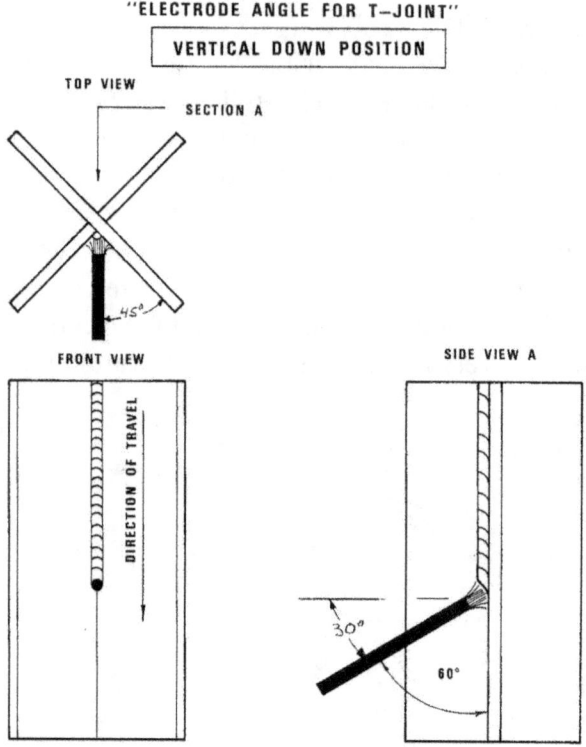

Figure 5-2 Electrode Work and Travel Angle For T-Joint

WELDING TECHNIQUE

You will learn how to use the drag (buried arc) technique to obtain full penetration V groove welds using AWS 6010 or 6011 electrodes. In addition, you will learn how to remove the deep slag lines associated with the drag technique.

The filler passes will be deposited using the downhill technique. On heavy wall thicknesses, the fill pass may be deposited using the downhill technique in the form of stringer passes. As an alternate, these passes can be deposited using the uphill technique. The uphill technique is easier to perform on heavy wall filler passes and results in a better weld appearance. It will be practiced in Chapter 6.

VISUAL INSPECTION

You have learned how to make welds in the vertical position by reading and understanding the techniques described in this chapter. After using the drag technique to deposit the first pass, you should check the root bead for incomplete penetration or excess penetration. The high crown on the drag stringer bead should be ground and the current setting increased in order to burn out the slag lines left from the drag stringer bead. The second pass should be concave and smoothly blended into each weld bevel. The fill and last cap passes should be inspected for external undercut, porosity, uniformity, and smoothness.

The purpose of the T -joint is to provide a quick weld joint, where you can obtain a lot of practice with a minimal amount of weld preparation. The T-joint provides four separate weld joints per each set up. Each joint can be welded using different techniques. In this chapter, all passes are deposited using an open arc, vertical down method. Each pass should be visually inspected after cleaning to observe and record non-uniform areas. You should record your reasons for the non-uniformity and review them with your instructor. Practice each technique until you are consistently making acceptable welds. Visual inspection is a very important tool for improving your welding skills.

NICK BREAK TEST

The nick break test is a fast economical method to check the internal weld metal for weld discontinuities such as lack of fusion, slag, and porosity. The corner weld is broken by using a hacksaw to saw 1/4 inch on each end. The weld sample is placed on a flat solid surface and hit with a two pound hammer until it is flat. Place in vise and hit one side until it bends then hit other side. Work the corner weld back and forth until it breaks. Observe the fracture surface for lack of fusion, porosity and slag.

PRACTICE T-JOINT VERTICAL DOWN (3F)

Make welds as described. Always clean each pass before depositing more weld metal. If the T-joint specimen exceeds 350 degree Fahrenheit, cool in water tank and dry before starting next pass. Inspect for weld discontinuities listed in Table 4-1.

PRACTICE CORNER JOINT VERTICAL DOWN (3G)

Make weld using technique described. All passes should be cleaned and inspected. If discontinuities are found, discuss with your instructor the possible causes and corrective measures.

Chapter 5 Review Questions

1. What is the name of the method used to complete the first pass of a 3G position corner joint when the direction of travel is started at the top of the plate?
2. What is the advantage of using the corner joint?
3. What is the joint design for an open root, vertical down weld using the drag technique?
4. What device can be used to check the root opening and land dimension?
5. During what circumstances can an AWS E-6010 electrode be used on straight polarity?
6. How should the arc length be adjusted from the 1G position to the 3G position?
7. In all arc welding, you should set the _____ on a piece of scrap before tacking the practice or production part.
8. Should the welding current be decreased when making the second pass over a drag root pass?
9. Is the work angle changed between the first pass and fourth pass when welding vertical down on the T-joint?
10. What will happen if the electrode is weaved too wide when filling the T-joint?

CHAPTER 6

ARC WELDING TECHNIQUES FOR THE VERTICAL UP (3G) POSITION

Learning Objectives

After mastering this chapter, you will be able to:

- Understand joint design and alignment for vertical up method
- Determine proper electrode size and arc length
- Learn to set welding current and control travel speed by observing weld puddle
- Control electrode angle and work angle when welding vertical up
- Weld V-grooves with backup in the vertical up position

In the last chapter you learned to support the weak liquid puddle with the force of the welding arc. When using the uphill welding method, the liquid puddle is supported by the solidified weld metal. That is, there is solid weld metal perpendicular to the pull of gravity. The weld is started by depositing a shelf of weld metal at the bottom of the T -joint and successive layers of weld metal are added until the weld pass is completed at the top of the joint.

All practice welds will be made using the T-joint. You will learn to deposit quality welds using this uphill method by mastering the welding techniques described in this chapter.

JOINT DESIGN

The practice welds will be made using a T-joint weld configuration. The plates are tack welded as shown in Chapter 4 under alignment. The T-joint provides four weld areas where each one represents a 45 degree weld groove on heavy wall (thick plate) without bevel preparation. The T -joint is one of the more practical weld practice set ups.

T-JOINT HOLDING FIXTURE

A holding fixture like that shown in Figure 6-1 may be used to hold the T-joint. The materials needed to fabricate the holding fixture are:

One 5-inch C-clamp

One 6-inch length of 2-inch angle iron

Two 4-inch lengths of ¾ -inch angle iron

This device can be used for the 1G, 3G and 4G positions by changing the orientation of the clamp on the welding table.

Figure 6-1 T-Joint holding fixture for 1G and 3G test positions

CLEANING

The 1½ " plate surfaces should be free of contaminates such as rust, scale, moisture, paint, dirt, and oil. An oxyacetylene torch can be used to remove moisture, paint, and oil. The rust can be removed with a pedestal grinder and sharp edges with a file. All weld passes shall be cleaned to remove slag and excessive weld deposits (knots). A chipping hammer and wire brush are most useful for removing slag. Files can be used to clean edges of weld area. The plates should be cleaned before alignment and tack welding.

ELECTRODE TYPE, SIZE, AND ARC LENGTH
Electrode Type

AWS E-6010 and E-6011 electrodes will be used to make the first practice welds. The direction of travel will be uphill instead of the downhill method used in Chapter 5. The E-6010 electrode will be used with DC reverse polarity and the E-6011 electrode will be used on AC current. The operating characters are similar; therefore, if you learn to weld with E-6010 you will not have any difficulty using E-6011. The choice depends on what type of current supply is available. In addition, the DC reverse polarity electrodes can be affected by arc blow; whereas, the AC electrodes are not affected.

The second type of electrode used will be the low hydrogen type such as E-7015, E-7016, and E-7018. The low hydrogen type of electrode must be deposited with a direction of travel vertical up. The liquid puddle is more fluid than a 10 series or 11 series electrode. In this chapter, the first two passes will be made using the E-6010 or E-6011 electrodes and the remaining passes will be low hydrogen type. Recall Figure 4-9, which shows the Lincoln Electric Excalibur 7018 electrode specifications.

Electrode Size

The electrode size for the E-6010 or E-6011 electrodes will be 1/8" diameter. For plate thickness less than 3/16", a 3/32" diameter may be used. The electrode size for the low hydrogen will be 3/32" for the first two passes and the remaining passes will be deposited with 1/8" diameter low hydrogen. The E-7018 electrodes have iron powder in the coatings; therefore, they deposit approximately the amount of weld metal one size larger. That is, 1/8 inch E-7018 diameter deposits the same amount as 5/32" E-6010 diameter.

Arc Length

When welding in the vertical up position, the arc length is shorter than that used for the vertical down position. The direction of travel is from the bottom of the weld joint toward the top. A solid weld metal shelf is established and the liquid puddle is supported by the solidified weld metal under the shelf. The welder always maintains solid weld metal between the weld puddle and the force of gravity.

The arc length for low hydrogen must always be kept shorter than that used for the 10 series or 11 series electrodes. The short arc length and increased fluidity is why the low hydrogen electrodes cannot be used in the vertical down direction.

WELDING CURRENT

Before making a production weld or practice test weld, set your welding current on a piece of scrap metal . Use the same diameter rod and place the scrap in the same position as the production part. The first try can be determined by one of the following methods:

1. Previous notes
2. Set by fluidity of puddle (puddle jumps if current high, puddle quiet if current low)

Welding current set for vertical up position is approximately 10 to 20 amps lower than vertical down position welding. After setting the current on scrap, adjust fluidity to that required to maintain a weld metal shelf. If the electrode is hard to start, the current is too low and if the puddle is jumping, the current is too high. The low hydrogen electrodes require more skill to strike (start) to avoid electrode sticking and starter porosity. It is particularly difficult to start low hydrogen on DC welding machines with low open circuit voltage or on AC welding machines. See Table 6-1 for suggested current setting.

TABLE 6-1 T-Joint Heat Settings for Vertical Up (3G) Position

Electrode Diameter	Plate Thickness	Thickness Decimal	Heat Amps	Puddle Surface
3/32	1/8	.125	105	jumping
3/32	1/8	.125	70	smooth
1/8	3/16	.188	135	jumping
1/8	3/16	.188	100	smooth
1/8	1/4	.250	135	jumping
1/8	1/4	.250	100	smooth
1/8	5/16	.312	135	jumping
1/8	5/16	.312	100	smooth

TRAVEL SPEED

The vertical up travel speed is controlled by observing the weld puddle shelf. A uniform weld pass is obtained by the smooth movement of the electrode. The arc is moved quickly across the middle of each weave and stopped at the edge of the puddle for approximately one second (dwell) as illustrated by solid dots in Figure 6-2. Liquid core wire metal is continually melting and flowing from the electrode into the puddle. Therefore, if the weld has excessive weld metal deposited, the electrode was held over that area too long.

As the electrode is moved from one side of the molten shelf to the other side, the electrode is moved upward approximately one electrode diameter. The liquid metal will flow to the hottest area; therefore, you should heat the sides more than the middle to prevent excess weld metal build-up in the center.

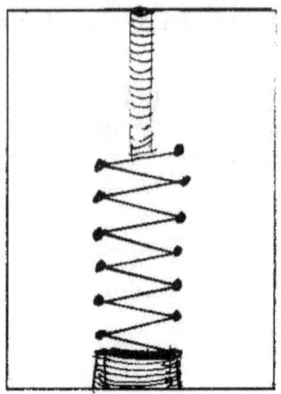

Figure 6-2 Electrode movement for
AWS E-6010 or E-6011 electrodes

ELECTRODE MOVEMENT

The electrode movement for vertical up position welding is a lower frequency, longer stroke (width) and is more precise than that used for vertical down. The uphill method is usually more difficult to learn than the downhill method.

The first pass uphill is deposited with a stepping motion. There is no side to side motion and the electrode is moved approximately two electrode diameters forward with an increase in the arc length. The electrode is returned to the puddle to deposit metal with slight dwell then moved quickly forward again. Figure 6-2 shows the illustrated electrode movement. This technique is only used for E-6010 or E-6011 type electrodes.

The electrode movement for the low hydrogen electrodes are shown in Figure 6-3. The arc length is held at constant short length and the electrode is stopped at each side momentarily (dwell). The travel angle is 10 degree push and the work angle is 90 degree. The electrode angles are shown in Figure 6-4

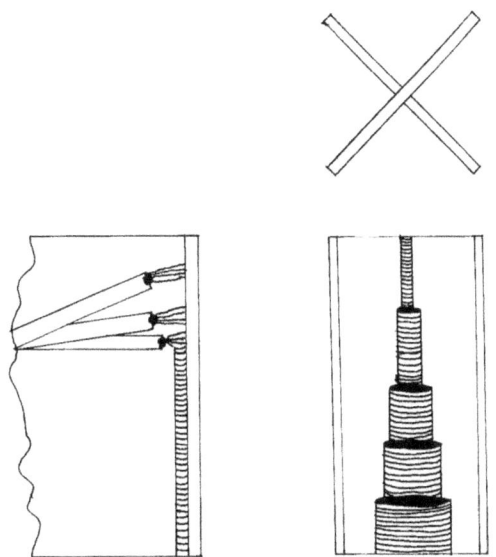

Figure 6-3 Electrode movement for AWS E-7018 low hydrogen electrodes

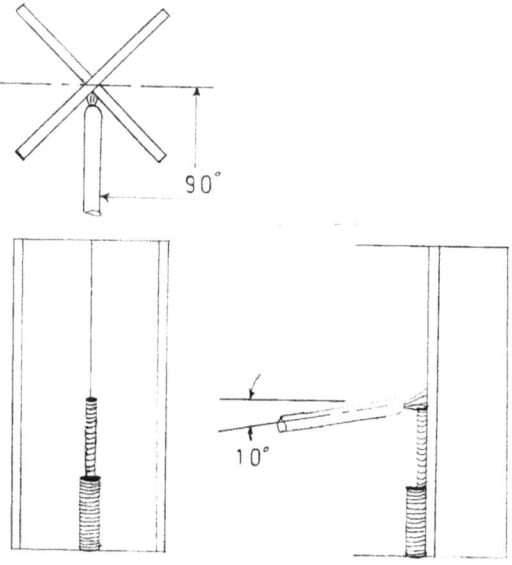

Figure 6-4 Work and Travel Angle for T-Joint

WELDING TECHNIQUE

You will learn how to deposit weld metal in the vertical up position using E-6010, E-6011, and low hydrogen type electrodes. Techniques will be shown on how to control the molten weld metal to prevent porosity and slag inclusions. The maintenance of the short arc length and smooth even electrode movement will be mastered after much practice.

The vertical up method is easier than the vertical down method for wide groove faces on heavy wall plates. Another advantage is less interpass cleaning is required due to the increase weld thickness per pass. The slower travel speed produces more heat input than the vertical down method which may be important when welding certain materials.

VISUAL INSPECTION AND PRACTICE

By using the T -joint, you can obtain many hours of practice with very little time spent on preparation of weld groove faces. You should make the first two passes using the E-6010 uphill method on all four weld grooves. Care should be taken not to overheat the T-joint sample. To minimize arc time, you may want to work with two T -joint samples, so that one can cool while you are welding on the other sample. If the heat input exceeds 350°F, the T-joint should be cooled with water, dried, and cleaned before continuing the welding operation.

Each pass should be checked for porosity, undercut, uniform weld reinforcement, incomplete fusion, and overall uniformity. The weld puddle is observed while welding and inspected after cleaning of each pass. As you learn to deposit each pass uniformly, the quality of the weld metal will improve.

If you desire to check the interior of the weld metal, one inch wide nick break coupons can be flame cut and broken as described in Chapter 13.

Chapter 6 Review Questions

1. What supports the liquid puddle when welding vertical up in the 3F position?

2. The low hydrogen type electrodes such as AWS E-7018 must be deposited in the _____ direction.
3. What are the sources of filler metal when using the AWS E-7018 type electrode?
4. When using the vertical up method, is the arc length longer or shorter than that used for the vertical down position?
5. Is the welding current higher for the vertical up position as compared to the vertical down?
6. Describe the electrode movement for vertical up welding when compared to vertical down.
7. Can the whipping motion be used for depositing low hydrogen weld passes?
8. Explain why the uphill method results in a thicker weld deposit.
9. Why is it harder to weld uphill than downhill for the first few passes?
10. Can AWS E-6011 be deposited using the uphill method?

CHAPTER 7

ARC WELDING TECHNIQUES FOR THE OVERHEAD (4G) POSITION

Learning Objectives

After mastering this chapter you will be able to:

- Select proper electrode size and arc length for overhead welding
- Learn to set welding current and travel control speed by observing weld puddle
- Control travel angle when welding overhead
- Weld V-grooves with backup in the overhead position

Very little welding is done in the overhead position because it is difficult to learn and the deposition rates are lower. If the production weldment cannot be turned, it may require welding in the overhead position. Examples are maintenance welding on heavy equipment such as trailers, tractors, and fixed position pipe welds.

All of the practice welds will be made using the T -joint and the holding fixture shown in Chapter 6.

SAFETY TIPS

The overhead welding position can be very dangerous since the weld spatter and slag will fall toward the welder. The following protective equipment must be used when overhead welding:

- Soft cotton hat to cover all hair
- Leather gloves
- Leather cape and sleeves
- Ear protection (ear plugs preferred)

Safety glasses with side shields under welding hood should be worn at all times (before, during, and after welding)

Slag may pop off of the weld during cooling; therefore, always wear eye protection when inspecting the weld after welding each pass.

When first learning to weld in the overhead position, it is advisable for two students to work together as a team. This provides an opportunity to observe and discuss welding techniques with one student welding and one student observing.

JOINT DESIGN

The joint design will be a 45 degree groove angle obtained by using the T-joint configuration. This simulates a 90 degree included angle which is larger than needed but provides quick practice coupons.

CLEANING

As with all welding, cleaning is very important. All plates should be free of rust, scale, moisture, coating, paint, dirt, and oil. The weld area should be ground or sanded before the plates are tack welded. All contaminates should be removed approximately ¼" to 1/2 " beyond the weld area.

ELECTRODE TYPE, SIZE, AND ARC LENGTH
Electrode Type

The AWS E-6010 or E-6011 electrodes will be used for all the overhead practice welds. This choice was made because a fast freeze electrode is easier to use in the overhead position. The E-6010 will be used with DC reverse polarity and the E-6011 will be used on AC current.

Electrode Size

All practice welds will be made using the 1/8" diameter. As the electrode size increases, the weld puddle will increase in size, making it more difficult to support against the pull of gravity.

After mastering the 1/8" diameter size in the overhead position, you may want to try the 5/32" diameter. Remember that the welding techniques learned on the 1/8" also apply to larger diameters, it just takes more practice to master the larger diameters when welding in the overhead position.

Arc Length

The arc length must be held very close when welding in the overhead position. If the arc length is not held close, the following will occur:

- Unstable metal transfer across arc resulting in more spatter
- Weld puddle size increases
- May lose support of weld puddle by arc force
- Arc force more forceful
- Arc voltage increases

It is necessary to maintain a much closer arc length in the overhead position than in any other position. That is why the overhead position is only attempted after all of the other welding positions are mastered.

WELDING CURRENT

Before welding on any production part or practice coupon, you should set the welding current on a piece of scrap metal. The current will be on the lower end of the workable range for overhead welding. After some practice, you will be able to set the current by holding a close arc and observing the weld puddle fluidity. This technique is very useful when welding with a new arc welding machine for the first time. You do not have to worry about the current lost in the welding cables and connectors since you are observing the weld puddle to determine the amount of current. The important parameter is the amount of current on the end of the electrode, not where the rheostat is set on the welding machine.

TRAVEL SPEED AND ELECTRODE MOVEMENT

The travel speed will be slower than the vertical down. For a right handed student, the direction of travel will be from left to right. The weld metal is deposited by using the stepping motion. The electrode is moved two diameters (1/4" if using 1/8" diameter electrode) forward quickly and metal is deposited for a distance of one and one-half diameters back toward the solidified weld metal. All weld metal is deposited in short one and one-half electrode diameter lengths across the overhead weld groove. As the weld groove widens, the deposit stroke is slanted across the bevels.

For bead width greater than two electrode diameter, the J technique should be used. The weld metal is deposited in a J movement toward the solidified metal, skipped forward on the other side and again deposited in a reverse J motion. This technique allows one side to cool while depositing metal on the other side, which results in a flat weld metal profile. See Figure 7 -1 for sketch of motion.

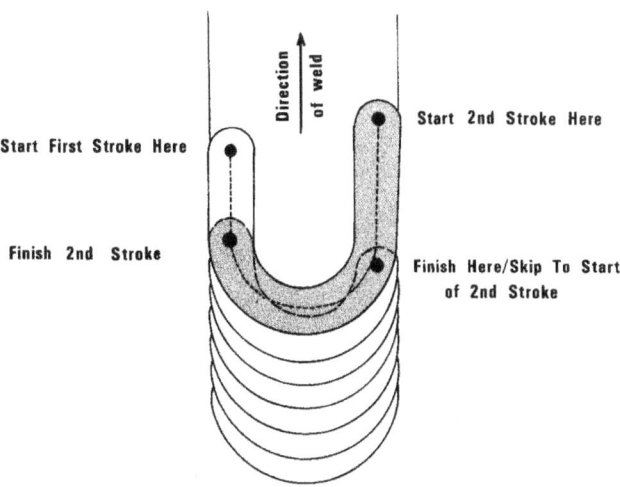

Figure 7-1 J-pattern electrode movement for overhead welding

ELECTRODE ANGLES

The T-joint is positioned as shown in Figure 7-2. The travel electrode angle is approximately 0 degree to 10 degree drag. The work angle is 90 degree measured from the imaginary plate surface. The work angle is an equal distance from each groove face.

Since metal is deposited directly off the end of the electrode, the travel and work angles are important. If the electrode is directed more toward the right groove than the left, there will be more weld metal on the right groove and the possibility of low fill and lack of fusion on the left groove face. If the travel angle is pushed instead of dragged, the weld metal fill will be rough and slag may be trapped.

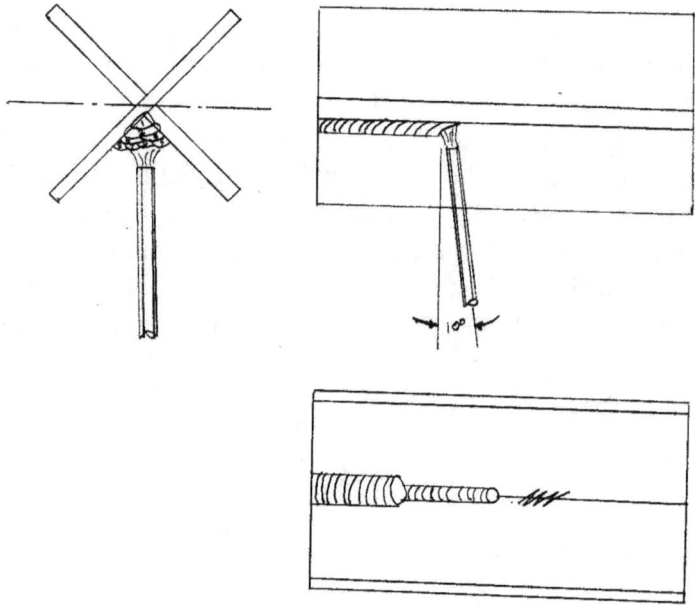

Figure 7-2 Work and travel angle for overhead position

WELDING TECHNIQUE

The technique used in overhead welding involves controlling the weak weld puddle by doing the following:

- Close arc length
- Short weld deposits

- Support puddle by pointing electrode perpendicular to liquid surface (balance force of gravity)
- Use lower end of current range

VISUAL INSPECTION

The surface of each weld pass is like a finger print. That is, it records any wrong moves that you make. Table 7-1 shows a list of weld discontinuities and causes.

It is always helpful to work with another student so that you can observe your partner's welding technique and talk to each other under your hoods. If you observe a wrong technique, tell your partner to make the correction. When starting to learn this position, your instructor should observe your techniques during welding and guide you into the correct arc length and electrode angles.

TABLE 7-1. Visual Inspection-Overhead (4G) Position

Fingerprint	Action Responsible
Spatter	long arc, high current
Undercut	long arc, no dwell
Misplaced bead	work angle less than 90 degrees
Bagged center	straight weave, no dwell on sides
Voids between ripples	irregular travel speed
Uneven beads	not observing puddle, wrong travel angle
Porosity	long arc, improper cleaning

PRACTICE

Make T-joint welds in the overhead (4F) position using the techniques described in this chapter. Do not start overhead welding until the safety tips are read, understood and implemented.

Chapter 7 Review Questions

1. Why is very little welding done in the overhead 4F position?
2. Why is the overhead position more dangerous?
3. Explain why the E-6010 electrodes can be operated in the overhead position.
4. List three reasons why the arc length is held close.
5. When welding in the overhead position, should the welding current be set higher than that used in the 3F position?
6. Is the travel speed faster or slower in the overhead position as compared to the 3F position?
7. What is the J technique?
8. What travel angle range is used for overhead welding?
9. List four techniques used to control the puddle in the overhead position.
10. What action is responsible for spatter?

CHAPTER 8

ARC WELDING
TECHNIQUES FOR PIPE

LEARNING OBJECTIVES

After mastering this chapter, you will be able to:

- Understand joint design and alignment for vertical down pipe welding
- Select proper electrode size for different pipe wall thickness
- Learn to set welding current and control travel speed by observing weld puddle
- Control electrode travel angles and electrode manipulation for vertical down pipe welding

The pipe welding chapters describe techniques that have been used since the 1930's to weld oil and gas lines. The excellent safety record of the pipeline welds in this country speak well of the techniques used by pipeline welders. An attempt has been made to share some of these techniques, so that a beginning student with reasonable hand eye coordination can master the art of Shielded Metal Arc (ARC) Welding of mild steel line pipe. If the American Welding Industry is to remain competitive, the time proven techniques must be shared so that quality and productivity can be maintained. Mastering these techniques will provide you an opportunity to earn a living as a certified pipeline welder.

JOINT DESIGN

The joint design shown in Figure 8-1 has been used since 1933 to arc weld oil and natural gas pipelines using the AWS E-6010 electrodes. The tolerances shown are not accurate enough to make quality pipeline welds. Pipeline welders speak in terms of root opening as a dime, penny, and nickel space. A dime thickness is equal to .050 inch thick and a penny is equal to .062 inch thick.

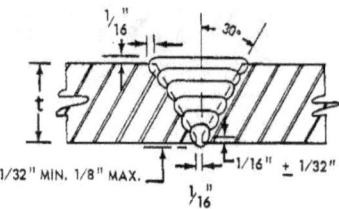

Figure 8-1 Joint Design for welding oil and natural gas pipelines.

You should use the same land (root face) dimensions as the root opening. If the root opening is a penny, you should file the land until a penny width is obtained. In the beginning, you may use a coin to measure or check the root opening and land but after a few days of practice, you will be setting the root and land with your eye.

The 30 degree pipe bevel is usually obtained with an oxyacetylene pipe beveling torch. The ring gear machine is split, so that it can be placed over a continuous welded pipeline to remove a section of pipeline. For training and qualification, a beveling machine such as shown in Figure 8-2 may be purchased to cut 6-inch pipe test nipples.

Figure 8-2 Pipe beveling machine in operation cutting six inch long, six inch diameter test nipples

After the bevel is flame cut and the land filed, the pipe inside surface should be filed or ground to bright, shiny metal approximately ¼" back from the pipe end. The outside surface should, also, be wire brushed to remove rust or scale next to the weld groove surface.

CLEANING AND GRINDING

The cleaner the pipe weld ends, the easier it is to make quality welds. Each pipe end should be mechanically cleaned to remove all paint, scale, rust, or dirt for a distance of ¼" on both the inside diameter and outside diameter. If the pipe has a submerged arc welded seam, the seam shall be ground flush with the inside surface. The pipe ends should be placed end to end to check for squareness. The high points should be marked for grinding until the two ends are parallel. The high points will have a heavy land due to the grinding; therefore, the bevel may require more grinding to form a true 30 degree bevel.

The result of using the drag technique for the first pass (stringer bead) will be a high crown or convex weld bead profile. The top of the crown should be removed with a portable power grinder. Any excessive weld buildup (arc starts) should, also, be removed. All tack welds should have the starts and stops ground.

The arc starts should be deposited in the weld groove faces. Excessive weld buildup on the fill passes should be ground to reform bevel to 30 degree angle. The start of the last pass (cap) shall be ground before the final tie-in is completed.

Each weld pass should have the slag removed with a hand wire brush or power brush before the next weld pass is started. The last pass should be wire brushed since a weld is not complete until it is properly cleaned. Also, it is impossible to inspect your weld unless it is properly cleaned. The cleaning should be done immediately so that any discontinuities can be marked and evaluated and, if needed, be removed.

ALIGNMENT AND TACK WELDING

It is important to align the pipe ends properly so that the test pipes have a common center line. The pipes must be held in place until the tack welds are completed. Several manufacturers make external line-up clamps and internal line-up clamps such as those shown in Figure 8-3. The internal line-up clamp shown was designed and fabricated by Cecil DeHaven, Columbia Gas Transmission welding instructor. Cross country pipelines are aligned by air assisted hydraulic clamps manufactured by CRC Evans.

Figure 8-3 Internal Pipe Clamp fabricated by Columbia gas transmission welding instructor Cecil DeHaven circa 1970

If line-up clamps are not available, the pipe segments can be placed in the vertical position and 1/16" shims can be used as root opening spacers. Since the first tack weld will close the root opening, you should set a root opening about a 1/32" wider than required for dragging the first pass. The tack welds should be limited to a maximum length of 1 inch to minimize the closing of the root opening. The second tack should be approximately180 degrees (opposite) from the first tack.

For small diameter pipe (less than 2"), two tack welds are sufficient. On pipe 4" in diameter or greater, you should use four tacks. After the first two tack welds, the space may be too tight. You should take a piece of broken hacksaw blade and grind a wedge shape on one end to drive open the tight space. The pipe ends can be hit with a hammer to close a space that is too wide. You should check the root opening using a penny or dime as a gauge before depositing any tacks.

You will make the tack welds using the same technique (drag) as the first pass because the tacks will become part of the first weld pass. The tack weld is made by striking the weld groove face with the end of the electrode to establish flowing current. Immediately push the electrode against the center of the weld groove to push the arc over the root opening and continually apply light pressure to the electrode. As the electrode burns, you should slide the electrode forward toward the direction of travel. Most of the light from the arc is inside the pipe and a small keyhole is formed under the electrode coating. After depositing about a *3/4"* long tack weld, you should flip the electrode out of the pipe joint with your wrist or your other hand. This prevents the keyhole from growing in size. If you let up on the pressure before flipping the electrode, the keyhole will increase in size resulting in a large hole. See Figure 8-4.

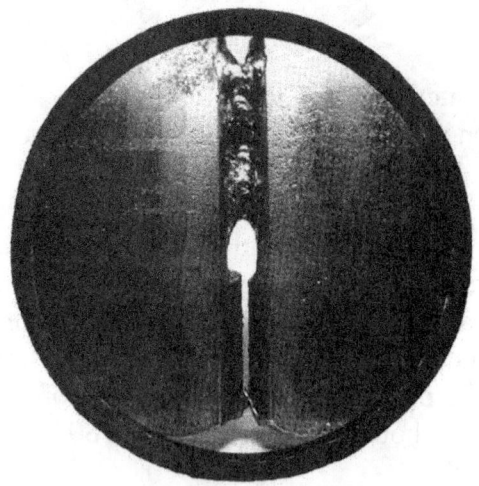

Figure 8-4 Large hole at end of tack caused by letting up pressure on electrode

The start of all tacks should be ground to remove the metal deposited before the electrode arc is buried. The tack start area is ground thin and tapered back into the body of the tack. The reason for the grinding is to provide a better chance to obtain complete fusion when the stringer bead is dragged into the tack weld. The objective is to obtain 100 percent fusion (full penetration) around the pipe circumference.

After tack welding, the pipe segment is placed in a pipe holding fixture. Most test holding fixtures are homemade. The fixtures are adjustable for the various fixed positions such as 2G, 5G, and 6G. The 1G or roll weld can be made on a rolling positioner with a joint roller.

FIRST PASS DRAG TECHNIQUE

Manual arc welding is still more economical than automatic welding machines due to the use of the drag technique on the first pass. The drag technique is fast and consistently produces excellent quality butt welds.

After the tack welded pipe is clamped in the horizontal (1G) position, you should start welding the first pass on the edge of a tack weld. The electrode is struck against the tack to establish an arc and pushed against the tack to start the drag technique. You should push down on the end of the electrode and forward towards the direction of travel (downhill). You will soon develop a touch (feel) as to the correct amount of pressure to apply on the electrode. You control the penetration by varying the amount of pressure (both forward and downward).

The drag stringer bead is continued from one tack to another tack. The drag stringer bead is stopped after the bead is fused into the next tack. When the drag technique is stopped on another solid part of the bead, the electrode is gradually raised to stop welding. The flip technique is only used when the bead is stopped in an open root.

The fine current control can be varied during welding of the drag stringer bead to increase penetration where the root opening is tight and to decrease penetration where the root opening is too wide. The current is changed either by the use of hand signals (field production) or a remote control rheostat. The penetration can, also, be increased by using a push electrode travel angle and decreased by using a drag electrode travel angle.

ELECTRODE TYPE, SIZE, AND ARC LENGTH

Electrode Type

The electrodes used for vertical down drag technique are AWS type E-6010, E-7010, or E-8010 made by several manufacturers. The first electrode manufactured for the drag technique was Lincoln Electric's Fleet weld 5 used by H.C. Price welders in 1933. If you master pipe welding techniques using a low strength electrode (E-6010), you will have no difficulty using the higher strength electrode of the same classification. Therefore, all practice welding can be done with E-6010 electrodes.

Electrode Size

The electrode diameter used for dragging the first pass is 3/32" diameter for wall thickness .140" to .156", 1/8" diameter for walls .188 to .249, and 5/32" diameter for wall thickness .250" and over. Refer to Table 8-1 for suggested electrode sizes. The beginning student should start with a 1/8" minimum electrode.

The electrode size, also, depends on the welding position of the pipe. A larger size electrode can be used for a 1G roll weld than for a 5G horizontal fixed pipe weld. A general rule of thumb is that the second pass (hot pass) should be run with the same diameter as the first (stringer bead) pass because the drag technique leaves two deep slag lines that must be melted by the second pass.

TABLE 8-1 Electrode Size and Number of Passes					
Wall Thickness	1st Pass	2nd Pass	Filler	Last Pass	Total Passes
.125	3/32"	1/8"	—-	1/8"	3
.188	1/8"	1/8"	—-	5/32"	3
.250	5/32"	5/32"	5/32"	3/16"	4
.375	5/32"	5/32"	5/32"	3/16"	4

Arc Length

In the preceding chapters, you have learned the importance of arc length in relationship to different weld positions and gravity. You should remember to look at the weld puddle and welding arc from a direction perpendicular to the electrode. When looking over the top of an electrode, it is nearly impossible to gauge the arc length. Therefore, never weld with your head looking over the electrode. When making a 5G or 6G fixed position weld, the arc length is short at the top and bottom positions and longer on the side positions. That is why pipe welding is more difficult than fixed position plate welding. You must gradually change from a flat to a vertical down to an overhead position while varying your techniques to overcome the force of gravity on the weak molten puddle.

WELDING CURRENT

The popular belief is that the qualified welding procedure tells you where to set the current. The broad ranges given in most welding procedures are of little value to a welder.

There is an easier and more reliable method to determine amperage. That is by observing the fluidity of the weld puddle. If the current setting is too low, the arc cannot be started. If the current setting is too high, the arc will be harsh resulting in a high arc force with spatter. Congratulations! You have just established the current range furnished by most welding procedures. Now, you must fine tune the current to meet your application.

A good analogy for fine tuning the current is adding paint thinner to a can of paint. If the paint is very thick, it is hard to spread and looks terrible. The same applies to welding current. If the current is too low, it is hard to start the arc and spread the puddle resulting in a terrible looking weld. If the current is increased so that the liquid puddle surface is jumping (vibrating), the puddle spreads and a concave smooth weld pass will be made with little effort by the welder. In addition, the chance of weld discontinuities is low.

Some factors that can affect the machine current settings are the length and size of the welding cables and the air temperature. The longer the leads and the smaller size cable, the more current is lost in the leads. The higher the cable temperature, the more current is lost in the cables. By using the puddle method, you do not need any calculations and the welding results are better and more reliable.

The welding current also depends on the pipe diameter, wall thickness, and pipe length. All of these factors can be controlled when the welding current is set by observing the puddle to obtain the right fluidity.

ELECTRODE ANGLES

One of the many variables that you must contend with is the angle of the electrode in relation to the pipe joint. When depositing the first pass using the drag technique, the electrode travel angle can be varied to either increase or decrease the amount of penetration. You can, also, control the keyhole size hidden under the electrode by changing the electrode angle. If the electrode is moved toward the direction of travel, the keyhole size will be decreased, whereas, if the electrode is moved toward the completed root pass (slight push travel angle), the keyhole size will be increased. This technique is very useful for preventing burn through where the root opening is too wide and for preventing incomplete root penetration where the root opening is too small or when approaching a tack weld.

You should always remember to position the end of the electrode (arc) perpendicular to the weld puddle. Since the coating is concentric to the core wire and the arc force is in alignment with the core wire, always point the electrode in the direction where joining is desired. For example, for pipe roll welds, the work angle is always 0 degree because the welder wants to fuse both weld grooves equally. The weld metal is deposited in the direction that the electrode is pointed.

TRAVEL SPEED

Most welding procedures indicate a travel speed range that the welder is required to use. From a practical viewpoint, these travel speed numbers are meaningless. How do you determine the travel speed while welding? It has been suggested to remember how fast you went on the weld before and if it was too slow, mentally record that speed and go a little faster on the next weld. This is easier said than practiced. The following paragraphs will describe an easier technique for controlling travel speed.

The travel speed technique that pipeline welders practice is the following. The welder monitors the appearance of the backside of the molten puddle and adjusts the travel speed so that the molten metal flows up against both grooves equally. You establish the amount of fill required for that weld pass and make slight travel speed adjustments to maintain a consistent fill level.

If the weld groove becomes wider or the previous pass is too low, you should slow the travel speed by monitoring the back side of the puddle. The result is that you can take non-uniform weld grooves and consistently make uniform, high quality weld passes. These puddle decisions are constantly being made during each welding pass and corrective measures are to be taken immediately.

ELECTRODE MANIPULATION

Electrode manipulation is very important for obtaining quality pipe welds. If you use the wrong electrode angles, you can practice for twenty years and still not make a quality weld. If you practice using the wrong electrode angles, you will show limited improvement. There are two things that you can count on: (1) the force of gravity is always present (2) the weld puddle can be controlled by using proper electrode angles.

From your first day of practice until your last, you will be learning to use proper electrode manipulation. The use of proper electrode manipulation techniques is necessary in order to obtain quality pipe welds.

VISUAL INSPECTION

You should always visually inspect every pass. If all passes are acceptable by visual inspection, the complete weld is likely to be free of weld defects. *As* described earlier, each weld pass is a finger print. Good moves as well as poor ones are visible in the completed weld passes. Some discontinuities to look for are surface porosity, external undercut, arc burns, excessive weld spatter, and uneven fill. You should ask why or how the discontinuity was formed and strive to eliminate them. Always remember to clean each weld pass before visual inspection.

WORKMANSHIP TIPS

Some useful workmanship ideas that right-handed pipeline welders have used:

- To prevent arc strikes on body of pipe, lay insulated side of electrode into groove before flipping hood down. Rotate the electrode until it is perpendicular for striking arc in groove.
- Hold electrode with left hand above weld location for starting arc. Lower hood and guide electrode into groove.
- Lightly touch right hand with left hand to steady right hand. Left arm resting on something solid, like pipe or vise grips.
- Hold electrode holder on end with light grip.
- Place new electrode in angle slot on holder to position electrode holder closer to pipe.
- Roll wrist as electrode burns to steady electrode and reduce strain.
- Plan ahead — reach for start location so that complete electrode is used before stopping arc.
- Bend electrode and rotate to angle desired.
- Use knee and elbow pads.
- Never leave electrode in holder in order to avoid accidental arc flash or chance of sitting on it.
- Maintain clean work area.
- Wear cotton cap.
- Wear ear protection.

Chapter 8 Review Questions

1. Why should each weld pass be cleaned before starting the next weld pass?
2. What is the purpose of the tack weld?
3. How can the penetration be controlled on the first pass when using the drag technique?

4. What technique can you use to obtain uniform width weld metal passes?
5. How is the weld puddle controlled with the electrode direction?
6. What is meant that each weld pass is like a finger print?
7. Why can you deposit more metal in the 1G roll position?

CHAPTER 9

PIPE WELDING TECHNIQUES FOR THE FLAT (lG) POSITION

༨༦༦༦༦

LEARNING OBJECTIVES

After mastering this chapter, you will be able to:

- Learn to set welding current and control travel speed by observing weld puddle in roll (lG) position
- Control electrode travel angles and electrode manipulation in roll (lG) position
- Weld full penetration V-groove pipe welds in roll position

The easiest pipe welding technique to learn is the roll (lG) position since the pull of gravity on the weld puddle is supported by the pipe when welding at the top (12 to 1 o'clock) position. When possible, all pipe welds should be made in the roll position because you can make quality welds with larger diameter welding electrodes (less time to complete weld). Consequently, the 1G pipe position is important to master. With practice, you will learn to maintain close control of the arc length and to deposit consistent and uniform weld metal by watching the weld puddle profile. These roll welding skills will be beneficial when you start practicing the fixed position welds.

TEST FIXTURE FOR 1G POSITION

All roll welds will be made using 6" diameter, .280" wall thickness pipe. Prepare the pipe ends as described in Chapter 8 and align the ends in a pipe clamp. Tack weld at four locations equally spaced around the pipe circumference to hold the 1/16" root opening. The practice test joint is secured in a 1G test fixture as shown in Figure 9-1.

Figure 9-1 1G testing fixture used to hold various pipe diameters Fabricated at Columbia Gas shop.

WELD BEAD SEQUENCE

All welding passes are completed by turning the pipe so that the welding arc is maintained at the 1 o'clock position. For the first pass, the pipe turner is locked with· the tack located at the top position. The drag technique is used to deposit a weld bead from the 12 o'clock tack to the 3 o'clock position. The brake is released and the pipe is turned to position the end of the drag bead (3 o'clock) at the 12 o'clock location. The first pass should be completed with four rotations of the pipe and all remaining weld passes will be made without using the brake.

The second pass (hot pass) is started and maintained at the 1 o'clock position by gently turning the pipe in an upward direction with your left hand. The welding current will be increased to burn out the slag lines left by the drag technique used on the first pass.

For current settings and control review Chapter 8. The sequence of welding passes is shown in Figure 9-2.

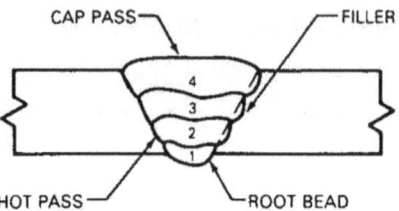

Figure 9-2 1G Welding Sequence

ELECTRODE SIZE AND ANGLES

A 5/32" E-6010 electrode will be used for all weld passes in the 1G position. A work angle of 0 degree is maintained unless an arc deflection (blow) is encountered, than the electrode holder is tilted toward the bevel where the arc is defected. If arc blow is experienced on the root bead, the electrode holder can be tilted toward the arc and given a quick twist (rotation).

The travel angle is 0 degree for a good joint fit up and is changed up to a 10 degree push travel angle for tight root openings (less than 1/16") and up to 10 degree drag travel angle for a wide root opening (more than 1/16"). The last pass (cap) is completed with a 0 degree work angle and a 0 degree travel angle. The end of the electrode should be moved horizontally across the weld groove to prevent long arcing on one side. By observing the puddle and concentrated practice, you will develop a smooth rhythmic motion, resulting in uniform weld caps.

VISUAL INSPECTION

Always clean each weld pass and observe the ripple patterns. If you are in a classroom situation, all irregular weld areas should be discussed with your instructor so that you can determine the cause of the discontinuous weld metal. Some possible causes are inconsistent travel speed, long or short arc length, wrong work or travel angle, too long of a pause at the weld start and too short of a pause at the end. If possible, you should work with another person, so that your electrode movements can be observed during welding and corrections can be made by talking to one another. The observer should position their head to observe the weld arc and the puddle from the same angle (perpendicular to arc) as the student welding.

PRACTICE

Practice making 6" diameter, .280" wall thickness, and 1G position (roll) welds.

Chapter 9 Review Questions

1. What are the benefits of welding in the 1G position?
   a. easier   b. maximum productivity   c. good quality   d. all
2. The filler passes are maintained at the 1 o'clock position by gently turning the pipe in an _____ direction with your non-welding hand.
3. A ___ travel angle is used to join tight root openings.
4. A ___ travel angle is used to join wide root openings.
5. Name five causes of uneven weld passes

CHAPTER 10

PIPE WELDING TECHNIQUES FOR THE HORIZONTAL (2G) POSITION

LEARNING OBJECTIVES

After mastering this chapter, you will be able to:

- Learn to set welding current and control travel speed by observing weld puddle in vertical (2G) position
- Control electrode travel angles and electrode manipulation in vertical (2G) position
- Weld full penetration V -groove pipe welds in vertical (2G) position

C hapter 10 illustrates techniques that you need to weld pipe in the vertical position. The practice pipes will be fixed with the pipe in the vertical direction and the weld groove in the horizontal position. The American Society of Mechanical Engineers (ASME) refer to this as a horizontal position, since ASME uses the weld groove as reference. The American Petroleum Institute (API) call this test position a vertical weld because the pipe axis is used as reference. To avoid confusion, you may want to use the American Welding Society (AWS) designation (2G) defined in Chapter 1.

If you are right handed, you must weld from left to right in the horizontal weld groove. The first pass will be welded using the drag technique explained in Chapter 8. The remaining passes will be deposited with the open arc technique.

The force of gravity (welders nemesis) will be pulling perpendicular to the fluid weld puddle; therefore, you will be balancing gravity with the force of the welding arc. The 2G fixed position is more difficult to master than the IG position.

TEST FIXTURE FOR 2G POSITION

Prepare the 6" diameter pipe as described in Chapter 8. Align the pipe ends in a pipe clamp and weld four tacks to maintain the 1/16" root opening. The test pipe is secured in a 2G holding fixture as shown in Figure 10-1.

Figure 10-1 2G Pipe holding Fixture

WELDING BEAD SEQUENCE

All weld passes are completed with the pipe fixed in the 2G position. The first pass is deposited using the drag technique. On the second pass (hot pass), the current is increased approximately 10 amperes and a tight arc length is used to burn out the slag lines remaining from the drag pass.

On the filler and cap passes, the current may need to be reduced approximately 10 amperes and a slightly longer arc length is used to support the liquid puddle. The welding bead sequence is illustrated in Figure 10-2.

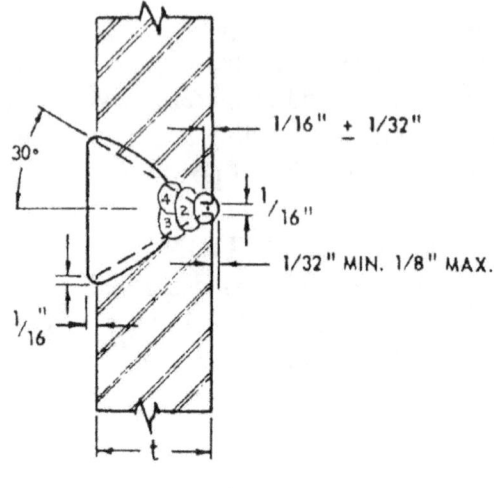

1/16" ± 1/32"

30°

1/16"

1/32" MIN. 1/8" MAX.

1/16"

t

t = LESS THAN 3/4"

Figure 10-2 2G Weld Sequence

ELECTRODE SIZE AND ANGLES

All of the 2G position welds are made using a 1/8" E- 6010 electrode. For all open arc weld beads, the work angle of the electrode is 10 degree toward the floor as shown in Figure 10-3. The work angle is zero degree for the drag bead.

Figure 10-3 The late Cecil DeHaven of Columbia Gas demonstrating the 2G work angle

The travel angle is zero degree for the drag bead and 20 degree for the open arc weld beads. Figure 10-4 illustrates the direction of travel and travel angle for the fill weld beads.

Figure 10-4 Welder demonstrating 2G travel angle

The cap (last pass) is completed using a series of overlapping weld beads called a "laced" technique. By using the laced technique, you can complete the 2G weld cap in one pass. The "laced" cap consists of a series of weld beads where weld metal is deposited from the top bevel to the bottom and you quickly slide the welding arc to the top bevel at a distance approximately one electrode diameter forward.

The electrode is momentarily stopped at the top start and at the bottom stop. After a few strokes (beads), you may want to gradually add a slight forward slope to the direction of travel. This slight angle will allow the slag to fall away from the weld puddle. Figure 10-5 shows the laced cap technique.

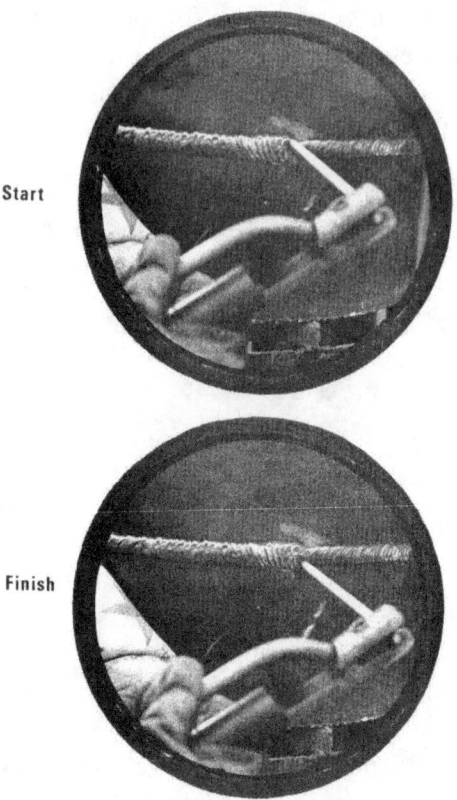

Figure 10-5 Laced cap technique used to weld the last pass on 16-inch diameter pipe in the 2G position

VISUAL INSPECTION
Always clean and inspect each pass. If the weld beads are irregular, determine the cause and make the necessary corrections.

PRACTICE
Practice 6" diameter, .156" wall, 2G position welds.

CHAPTER 10 REVIEW QUESTIONS

1. Why is the 2G position more difficult to learn than the IG position?
2. Why is the current increased on the second weld pass?
3. How is the arc length used to control the weld puddle?
4. Why is the work angle different for the first pass than the filler passes?
5. What technique is used for the cap pass?
6. Why is each weld pass cleaned?
7. What advantage does the laced technique have over the weave?

Chapter 11

PIPE WELDING TECHNIQUES FOR THE HORIZONTAL FIXED (5G) POSITION

Learning Objectives

After mastering this chapter, you will be able to:

- Learn to set welding current and control travel speed by observing weld puddle in horizontal (5G) position
- Control electrode travel angles and electrode manipulation in horizontal (5G) position
- Weld full penetration V-groove pipe welds in horizontal(5G) position

This chapter describes techniques that you need to weld when the pipe is in the horizontal fixed position. The first pass will be welded with the drag technique and the remaining passes will be completed with the open arc techniques.

The force of gravity (welders nemesis) will be pulling the weld puddle toward the travel direction (down); therefore, you will be balancing gravity with the electrode arc force. The 5G (bell hole) fixed position weld is more difficult to master than the 2G position due to the continuous change in weld puddle position from flat, vertical down, to overhead.

TEST FIXTURE FOR 5G POSITION

Prepare the 6" diameter pipe as described in Chapter 8. After the pipe ends are prepared, aligned and spaced *(1/16")*, four tacks are made with one at each quarter section. The tack welded test pipe is secured in a 5G test fixture as shown in Figure 11-1.

Figure 11-1 The late Cecil DeHaven of Columbia Gas observes the weld puddle while a student learns to weld in the 5G position. The student could breathe less fumes if he did not lean over the arc plume.

WELDING BEAD SEQUENCE

The first pass is deposited with the drag technique. The electrode is placed in the holder using the 45 degree slot (slanted toward hand). You should grip the electrode holder at the slotted end since the drag technique requires pressure on the electrode. Figure 11-2 shows a pipeline welder using the drag technique.

Figure 11-2 Welding student uses the drag technique to deposit the first pass. When properly executed, the drag technique arc and fume is internal to the pipe.

In the filler and cap passes, the current is increased approximately 10 amperes from that used on the drag pass. The arc length is held close at the 12 to 1 o'clock and the 5 to 6 o'clock positions and gradually lengthen for the 2 to 4 o'clock positions. The weld bead sequence is shown in Figure 11-3.

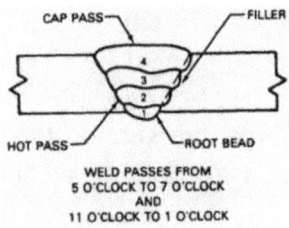

Figure 11-3 5G Welding Sequence

ELECTRODE SIZE AND ANGLES

All of the 5G position welds are made using a 5/32" E-6010 electrode. The electrode work angle is zero degree, that is the electrode is 90 degree from each pipe surface. The electrode travel angle varies as the welder progresses from the top to the bottom of the pipe. The electrode travel angle is zero degree at 12 o'clock, increases to 30 degree drag at 3 o'clock, and decreases to zero degree at the 6 o'clock position.

The cap pass is difficult to keep flat in the overhead (6 o'clock) position. A trick often used by pipeline welders is the J technique. This technique consists of a quick movement forward on one bevel, then deposit weld metal back toward the completed weld for a short distance ending with a hook in the puddle. The distance is approximately two electrode diameters. You make a quick move forward three electrode diameters on the opposite bevel and deposit weld metal toward the puddle again hooking the end into the puddle. By timing these moves, the completed cap will be flat. The technique is illustrated in Figure 11-4.

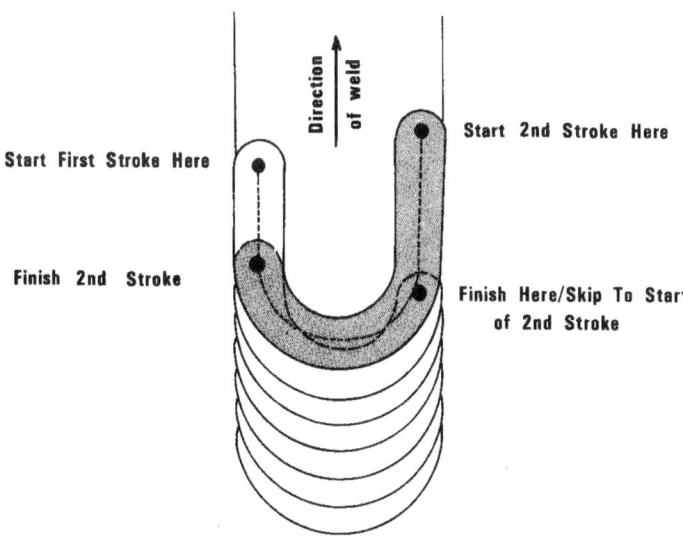

Figure 11-4 Electrode movement for overhead (six o'clock) position

VISUAL INSPECTION

The fill passes should have a concave surface with smooth continuous surface. The cap pass should be uniform with a maximum of 1/8" buildup beyond the pipe outside surface.

PRACTICE

Practice 6" diameter, .250" wall, 5G position.

CHAPTER 11 REVIEW QUESTIONS

1. Why is the 5G position harder to learn than the 2G position?
2. How many tacks are required to hold the root opening (space) for the 6" diameter pipe?
3. When using the drag method, why should you grip the electrode holder at the slotted ended?
4. Explain why the arc length is increased from the 2 to 4 o'clock position and the 10 to 8 o'clock position.
5. In the 5G fixed position, the work angle is _____ degree.
6. Why is the travel angle changed on the sides of the pipe?
7. How can a welder control the weld puddle at the 6 o'clock position to obtain a uniform cap pass when welding in the 5G position?
8. After completing three passes, why is the pipe groove filled more at the top and bottom positions?
9. What is the advantage of using 5/32" diameter electrode over 1/8" diameter?
10. What is the slang name for a 5G pipe weld position?

Chapter 12

PIPE WELDING TECHNIQUES FOR THE 45° FIXED (6G) POSITION

LEARNING OBJECTIVES
After mastering this chapter, you will be able to:

- Learn to set welding current and control travel speed by observing weld puddle in a 45 degree (6G) position
- Control electrode travel angles and electrode manipulation in a 45 degree (6G) position
- Weld full penetration V-groove pipe welds in the 45 degree (6G) position

This chapter covers the techniques needed to weld pipe that is positioned 45 degree from Level. The 45 degree (6G) test weld is used to qualify welders who construct pipelines in hilly terrain since some cross country pipelines have inclines that approach 45 degree. One of the advantages of the drag technique is that the pull of gravity does not affect the keyhole puddle.

The force of gravity will affect the open arc weld puddles; therefore, you will learn techniques to balance the gravity affect. The 6G fixed position is the most difficult weld test position to master.

HOLDING FIXTURES FOR 6G POSITION

You should prepare, space, and tack weld the pipe sections as shown in Chapter 8. The tack welded pipe sections are secured in a 6G test fixture with the widest root openings placed at the 3 and 9 o'clock positions. Figure 12-1 shows a pipeline welder completing a 6G test weld.

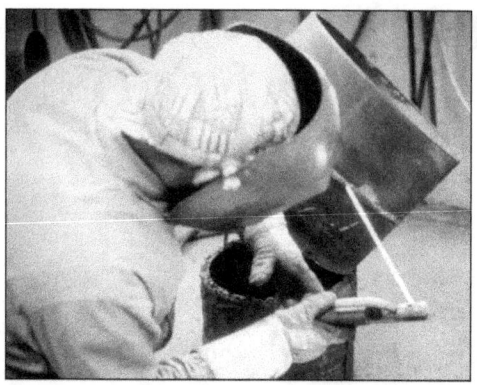

Figure 12-1 6G Test Weld. Keep the molten metal puddle level with the floor even though the pipe sits at a 45 degree angle.

WELD BEAD SEQUENCE

The first pass is completed using the drag technique as discussed in Chapter 8. The welding current and arc length are similar to the techniques used for the 5G fixed position.

ELECTRODE SIZE AND ANGLES

All of the weld passes will be made with 5/32" E-6010 electrodes. The force of gravity affects the liquid weld puddle at a compound angle since the pipe axis is at 45 degree from the floor. The electrode work angle is more complex than 5G due to the effect of gravity. The electrode work angle will vary from zero degree at the top to 10 degree on the side· and return to zero degree on the bottom.

The electrode travel angle varies as the welder progresses from the top to the bottom of the pipe. The electrode travel angle is zero degree at the top, increases to 30 degree drag on the sides and decreases to zero degree on the bottom.

The direction of electrode movement across the puddle (oscillation) is not from bevel to bevel as in the 5G position. The top first stroke is bevel to bevel and gradually moved to a slight lead. The direction of movement across the puddle is level to the floor. If water is poured into the pipe, the electrode is moved across the bevel at the water level that is parallel to the level floor. Figure 12-2 shows close-up view of the weld ripple patterns at the top, side and bottom positions.

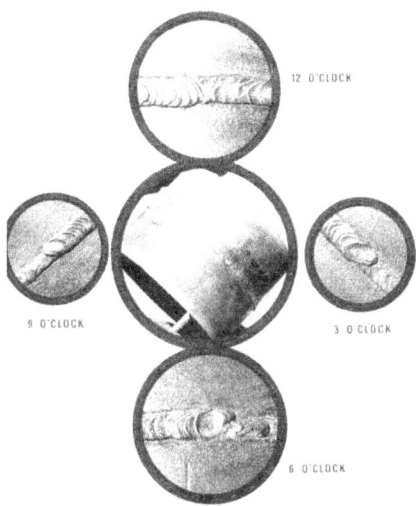

Figure 12-2 Close up of the ripple patterns at the top, side, and bottom locations for a 6G position weld.

DESTRUCTIVE TESTS

Figure 12-3 shows a properly made 6G test weld. The test weld is being destructive tested per API 1104 Standard. The instructor is cutting 4 nick break coupons, 4 root bend coupons, 4 tensile coupons, and 4 face bend coupons for performance qualification testing of a contractor welder. The most difficult test to pass is the nick break test as explained in Chapter 13.

Figure 12-3 Instructor Cutting Coupons for testing six-inch diameter pipe.

PRACTICE

Practice 6" diameter, .280" wall thickness, 6G test position.

CHAPTER 12 REVIEW QUESTIONS

1. Why is the 6G fixed position the most difficult weld position to learn?
2. What clock position should the widest root opening be placed?
3. The work angle is _____ degree at the 3 o'clock location when welding in the 6G position.
4. The travel angle is _____ degree at the 3 o'clock position for a pipe weld made in the 6G position.
5. Is the electrode movement across the puddle from bevel to bevel?

CHAPTER 13

NICK BREAK TEST

LEARNING OBJECTIVES
After mastering this chapter, you will be able to:

- Evaluate weld quality with the nick break test
- Explain how the nick break test works
- Prepare and conduct a nick break test
- Interpret nick break coupons
- Identify defects such as slag, porosity, incomplete fusion, inadequate penetration, and cracks

The nick break test is the most important destructive test and it is used to evaluate welding procedures and welder performance. It can also be used to verify nondestructive testing results such as radiographic interpretation of butt welds. A properly prepared and broken nick break test can determine if the image on the radiographic is a crack or a slag line. The nick break test is a very cost effective method requiring only a hacksaw, cutting torch, and a hammer.

The nick break test can be used to test butt welds as well as fillet welds, lap welds, and full penetration branch connection pipe welds.

Development of the Nick Break Test

One of the earliest users of the nick break test was Columbia Engineering Corporation in the early 1930's. The Columbia System still uses parts of this code to measure welder performance today.

The only national welding code that uses the nick break test is the American Petroleum Institute Standard 1104. The first edition was published in 1953. The welding code developed in the early 1930's by Columbia's welding engineer D. O. Ferguson preceded welding codes by over twenty years. The first illustration of electric arc welding nick breaks was published by the American Welding Society as an appendix B of ANSI / *AWS* D10.12-89, "Recommended Practices and Procedures for Welding Low Carbon Steel Pipe". The 1950 American Welding Society Welding Handbook contained specification on the nick break test.

THEORY OF THE NICK BREAK TEST

The nick break test is designed to break through the weld metal instead of the base metal. This fracture is accomplished by sawing a notch in the weld metal, and cold working the coupon in two directions until a crack propagates through the weld metal and the coupon breaks. The fracture path will break through the discontinuities in the weld metal, if any exist. The exposed discontinuities are visually evaluated and compared to a predetermined standard of acceptability.

The nick break test is a very practical test to open up the weld deposit for visual examination. A welder can leave small discontinuities in a bend test and/or tensile test which may not open to the surface. All welding discontinuities may not be detected with non-destructive testing or they could be misinterpreted. The nick break test, if done properly, will always break through the discontinuities.

Some codes permit bending or pulling the nick break coupons in a tensile machine to save time. If the tensile machine is used to break the nick break coupons, the fracture may travel from the notch into the base metal. Because the nick break coupon did not break through the weld, the test will not reveal any weld metal discontinuities. This invalidates the nick break test since the purpose is to expose any discontinuities found in the weld metal.

COUPONS PREPARATION

The nick break coupons for butt welds are prepared by flame cutting straps perpendicular to the weld with a minimum of 3" on each side of the weld and a width of approximately 1". The 1" width is reduced to 3/4" by sawing a 1/8" deep notch with a hacksaw (or handsaw). For high strength weld metal, the weld reinforcement may be notched approximately 1/16" deep across the width connecting the 1/8" notches. Figure 13-1 shows an illustration of a nick break coupon for a butt weld test.

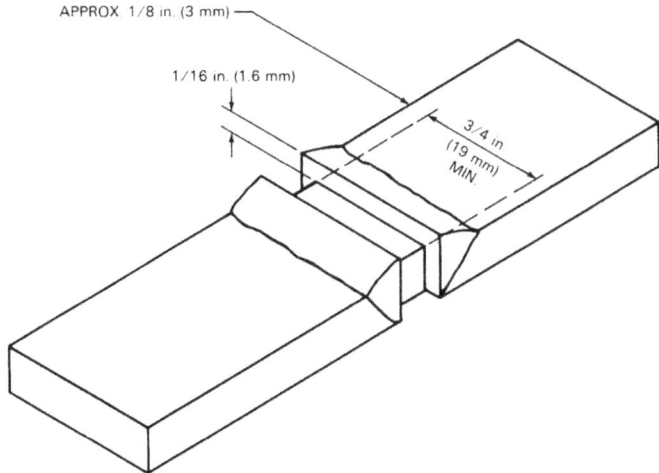

Figure 13-1 Butt weld nick break specimen.

The nick break test can, also, be used to evaluate fillet welded joints. Fillet welded lap joints are tested by flame cutting a 3" wide by 6" long coupon. A notch is sawed 1" long on each end which leaves a 1" center for evaluation. Figure 13-2 shows a lap weld that is notched and ready for breaking.

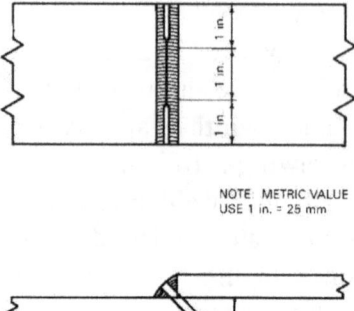

Figure 13-2 Lap weld nick break specimen.

A pipe sleeve nick break coupon is prepared by flame cutting the 1" notch and a saw cut across the top of the 1" middle section. The details are illustrated in Figure 13-3.

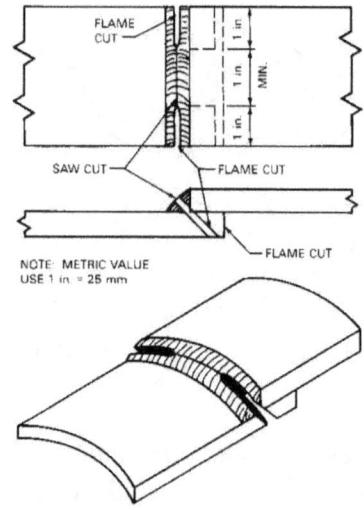

Figure 13-3 Sleeve nick break specimen.

Pipe branch connections can, also, be evaluated with the nick break test. Two nick break coupons are flame cut from the pipe branch crotch and two test coupons are flame cut from the top and bottom points. The ends are flame cut and the weld reinforcement is sawed about 1/8" deep. Properly prepared coupons for branch connections are shown in Figure 13-4.

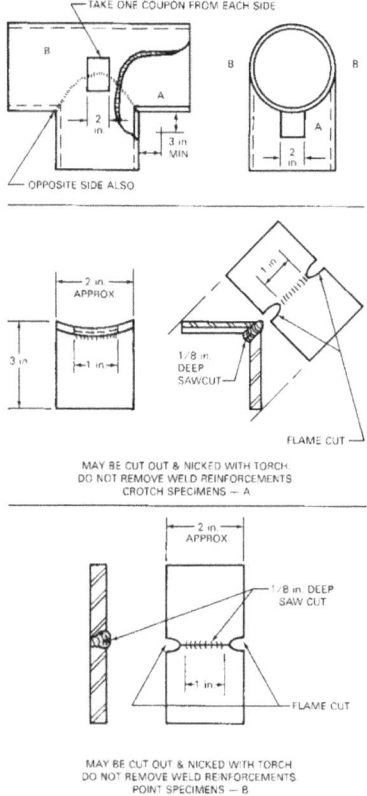

Figure 13-4 Branch nick break specimen.

TESTING PROCEDURE

All nick break coupons are always broken by supporting the ends and hitting the notched weld with a hammer. The notch is closed with one or more strikes then reversed 180 degrees and the second notch is closed. The coupon is continually hit, rotated, and hit until the coupon breaks.

The fractured surfaces are examined for evidence of weld discontinuities such as porosity, incomplete penetration, slag inclusions, incomplete fusion, and cracks. After identification, the discontinuities are evaluated in accordance with the acceptance standards established in the fabrication document, or by the facility doing the training or qualification.

A sample acceptance standard for welder training is given in Table 13-1. A good welding instructor should be able to give the welding student advice on how to eliminate all welding discontinuities. Some examples of causes of weld discontinuities are improper welding technique, improper cleaning, poor bevel preparation, wrong travel speed, and wrong current setting.

Table 13-1 Typical Nick Break Acceptance Standards	
Type of Discontinuity	**Requirements**
Inadequate penetration	none
Incomplete fusion	none
Porosity	1/16" maximum
Slag inclusions	1/8" maximum length
Cracks	none

TEST FIXTURE

The test fixture is used to provide support for the ends of the nick break coupons while the weld area is hit with a hammer (weight 1 to 8 lbs) so that the notch will propagate through any discontinuities in the weld metal. A simple test fixture can be fabricated from 6" diameter pipe as shown in Figure 13-5. The pipe is slotted on the center line and the material 90 degree from the slots is contoured to provide clearance for the hammer handle.

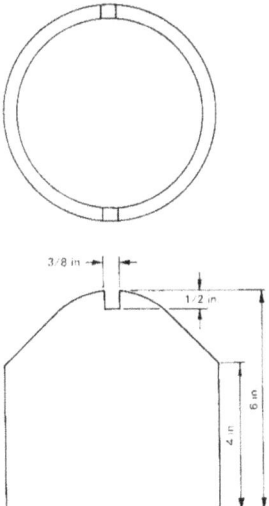

Figure 13-5 Easy-to-make nick break tester.

A vise can be utilized to hold the coupons but it requires more time. In addition, the broken coupon may hit someone as it flies across the room.

The fillet weld coupons are more difficult to hold and break. A blacksmith anvil may be used to break nick coupons as shown in Figure 13-6. The test coupon is held with tongs or pliers on the anvil and hardy tool. The coupon is hit on the weld two times then turned over and hit two times. The sequence is continued until the nick coupon is broken. The crotch coupon is placed on the anvil and hit as shown in Figure 14-6.

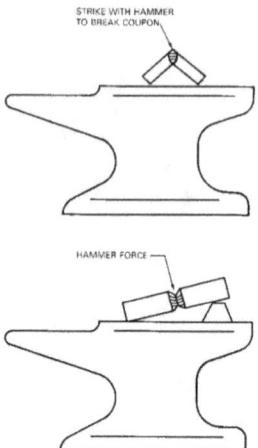

Figure 13-6 Nick break coupons can
be broken on a blacksmith's anvil

INTERPRETATION

The reading or interpretation of nick break specimen is more difficult than other destructive weld metal tests because the fractured surface must be observed to determine which ones are actual defects, and which are not.

Weld metal discontinuities are stress risers. Therefore, the weld metal fractures from the sawed notch through any of the discontinuities that are present. Some of the more common weld defects are described in the following paragraphs.

Slag inclusions

The American Welding Society (AWS) defines **slag inclusion** as "non-metallic solid material entrapped in weld metal or between weld metal and base metal". The nick break fracture will travel from the cut notch to the slag inclusion and through the center of the inclusion. Therefore, the slag inclusion is visible on both fractured surfaces of the nick break.

The slag inclusion may have a black glass-like appearance, or it may have a smooth yellow colored contoured surface. The location of the slag inclusion is sometimes smooth because the slag has been dislodged by the force of the hammer blows breaking the specimen.

It is useful to match the two specimens together and rotate the specimen in good light. Observe both fractured specimens at the defect area. Sometimes it is easier to read one fracture surface than the other matching side. One side may be smooth while the other side may still have slag trapped on the fractured surface. Figure 13-7 shows an elongated slag inclusion.

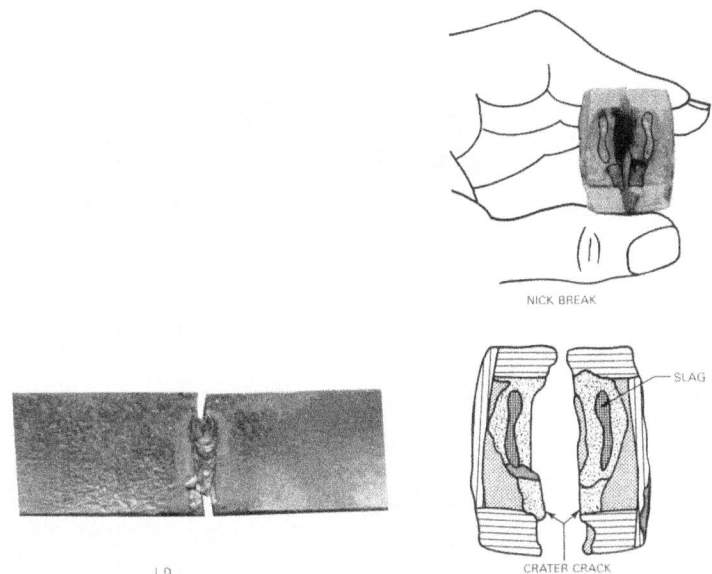

Figure 13-7 Fracture surface exposing elongated slag and crater crack.

Porosity

AWS defines **porosity** as "cavity type discontinuities formed by gas entrapment during solidification". Porosity type defects are always spherical in shape and may be isolated or grouped in clusters. The key to the identification of porosity is the spherical shape and absence of non-metallic solid material. Porosity has a bright white or silvery appearance if it is not open to the atmosphere. Surface connected porosity usually has a black oxide appearance. The defect free weld metal fracture surface has a gray color without voids. An example of porosity is shown in Figure 13-8.

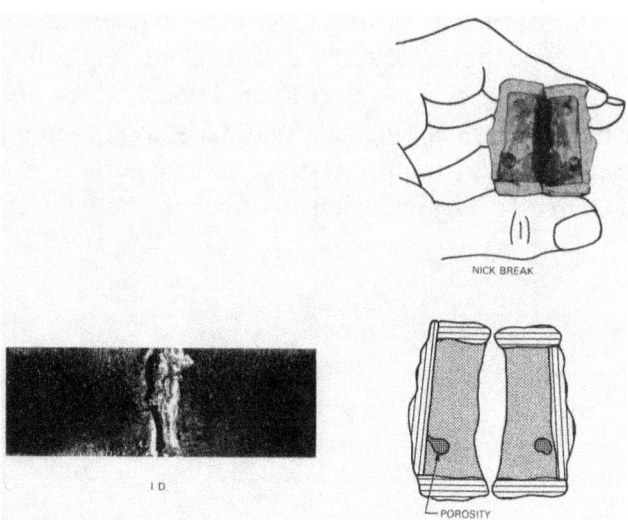

Figure 13-8 Fracture surface with porosity (gas pockets).

Incomplete fusion

API standard 1104 defines **incomplete fusion** as "the lack of bond between weld beads or between weld metal and the base metal". Incomplete fusion results from a solid surface not being melted (welded), commonly called "no weld" condition. Therefore, the shape of the defect depends on the weld joint design. If the joint design is a single vee, the base metal incomplete fusion would be a planar type shape showing the area of the bevel surface which is not fused.

In some cases, the grind marks on the original bevel can be identified. It is helpful to match the two broken nick break specimens together and identify the location of the first weld pass and the last, as well as the weld bevel area. If the defect is located on the bevel surface or between weld passes and is planar in shape, it could be incomplete fusion.

Sometimes the planar surface defect extends beyond the fractured surface showing a laminar type appearance. The incomplete fusion defects are shown in Figure 13-9. The incomplete fusion area will be brownish to black in color as compared to the gray weld metal that is clear of defects.

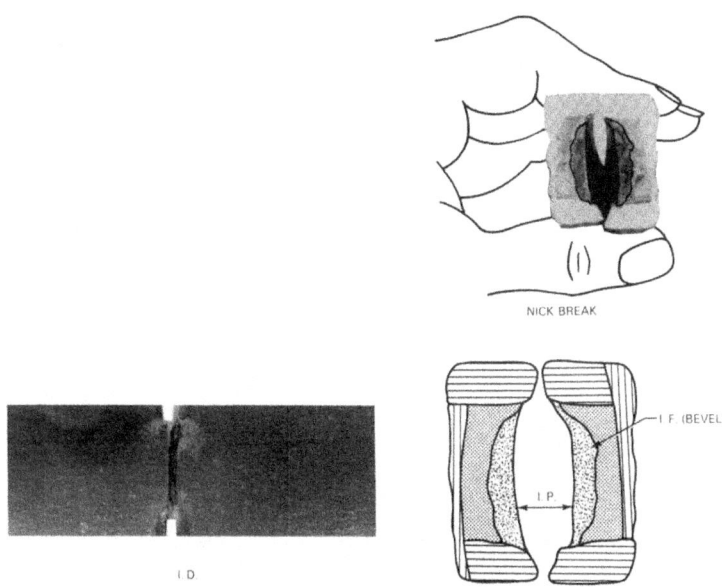

Figure 13-9 Incomplete fusion fracture surface

Inadequate Joint Penetration

API 1104 defines **inadequate penetration** as "the incomplete filling of the weld root with weld metal". The fracture path will extend from the sawed notch through sound weld metal to the inadequate joint penetration area. The defect is easy to identify in that it is always located at the weld root area and it is planar in shape. An inadequate joint penetration can be detected in the nick break specimen before it is broken. The nick break specimen will show how deep the inadequate penetration extends into the weld metal. The inadequate joint penetration is black to bluish in color. An example is shown in Figure 13-10.

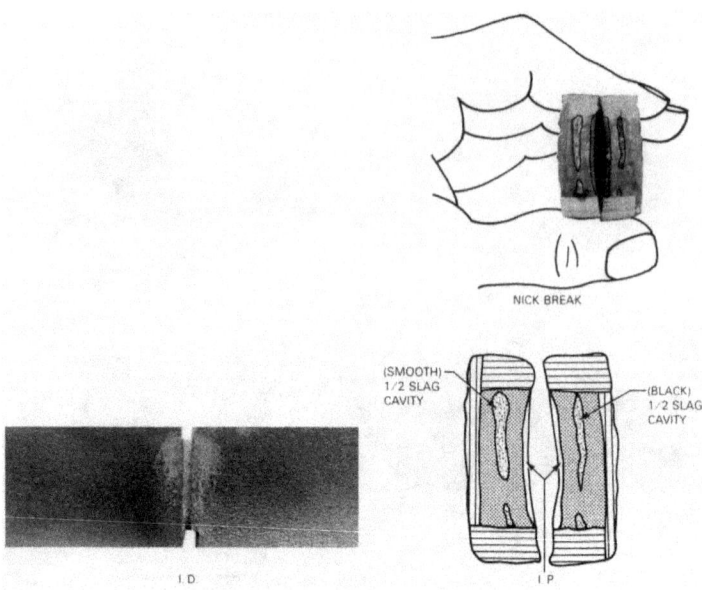

Figure 13-10 Incomplete penetration fracture surface.

Cracks

AWS defines a **crack** as, "a fracture type discontinuity characterized by a sharp tip and high ratio of length and width to opening displacement". A crack is a discontinuity located in the weld metal or base metal. The crack fracture surface is flat and is silvery in color if the crack occurs after the welding is completed. If the fractured surface of the crack shows a blue oxide color, the metal cracked before the final weld passes were completed. The crack surface was heated to the temper color range by later weld passes. A cracked weld nick fracture is shown in Figure 13-11.

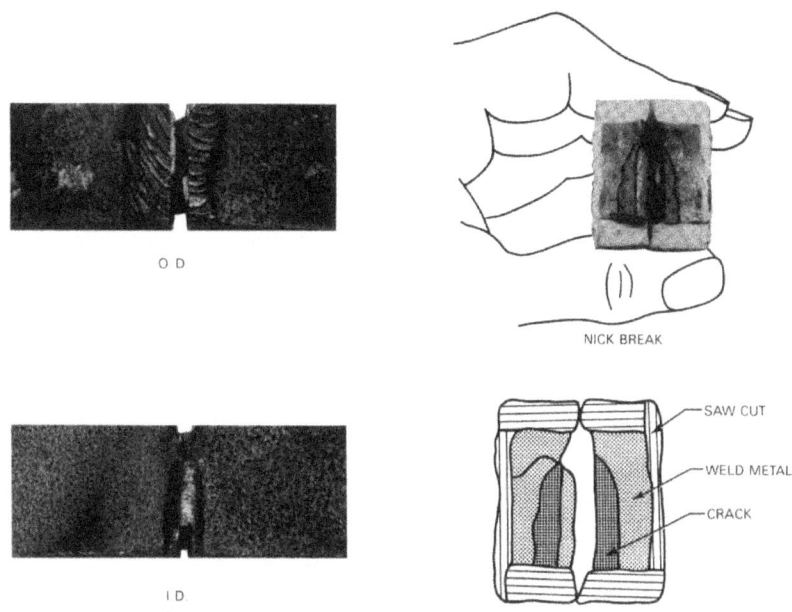

Figure 13-11 Crack fracture surface.

Chapter 13 Review Questions

1. Name three characteristics of the nick break test that make it unique.
2. What national standard specifies the use of the nick break test?
3. Explain the difference between a weld metal bend test and a nick break test.
4. Why is a hammer used to break nick break coupons?
5. Sketch the test fixture for breaking a butt weld.
6. When does a discontinuity become a defect?
7. Why does the fracture travel from the sawed notch through the weld defect?
8. Describe a slag inclusion.
9. State the difference between porosity and incomplete fusion.
10. Inadequate penetration is visible from the. _____ plate surface.

11. If the first weld pass cracks before the last pass is completed, what color is the flat crack surface?
12. What is the most severe weld discontinuity?
13. What is the least severe weld discontinuity?
14. Explain the difference between weld discontinuity and weld defect.
15. What tools are required to perform the nick break test?

Science of Personal Achievement

W e will be studying the seventeen principles of success that Andrew Carnegie disclosed to Napoleon Hill almost one hundred years ago. These principles tell how any person may learn to use the secrets of America's greatest and richest man. Andrew Carnegie gave the world steel at a reasonable price and used his millions of dollars of profit to establish learning centers throughout the world. This seminar is based on the book titled How to Raise Your Own Salary by Napoleon Hill which gives the verbatim success talks he had with his sponsor Andrew Carnegie.

Napoleon Hill recorded Andrew Carnegie's answers to questions concerning his seventeen steps to success. Napoleon Hill states "that the quality of service you render, plus the quantity, plus the mental attitude, in which you render it, determines the sort of job you hold and the amount of pay you receive." I have included several excerpts from this interview below which highlight the main ideas behind the seventeen steps.

Principle 1 Set a Goal: Purpose

Ten qualities of personal power are: (1) The habit of definiteness of purpose (2) Promptness of decision (3) Soundness of character (intentional honesty) (4) Strict discipline over one's emotions (5) Obsessive desire to render useful service (6) Thorough knowledge of one's occupation (7) Tolerance of all subjects (8) Loyalty to one's personal associates and faith in a Supreme Being (9) Enduring thirst for knowledge (10) Alertness of imagination

Principle 2 Associate with People Having Like Interests

Defined by Carnegie: "An alliance of two or more minds, working together in the spirit of perfect harmony, for the attainment of a definite purpose." The nine motives behind everything we do are: (1) The emotion of love (2) The emotion of sex (3) Desire for financial gain (4) Desire for self-preservation (5) Desire for freedom of body and mind (6) Desire for self-expression (7) Desire for perpetuation of life after death (8) The emotion of anger, often expressed as envy or jealousy (9) The emotion of fear

Principle 3 Develop an Attractive Personality

Traits of an Attractive Personality are: (1) Positive mental attitude (2) Promptness of decision (3) Tone of voice (4) The habit of smiling (5) Tolerance (6) Keen sense of humor (7) Faith in infinite intelligence (8) Genuine fondness for people (9) Humility of the heart (10) Personal magnetism

Principle 4 Employ Applied Faith

Applied faith gives power to all who apply it. It is the great equalizing force which truly makes all men equal. The power that removes all limitations from the mind is Faith; and I have emphasized the fact that the source of all Faith is belief in Infinite Intelligence. Once you understand this truth, you will not need to worry about self-confidence.

What knowledge I have of the power of the mind, I acquired from the greatest of all schools, the University of Life, I have one other suggestion to offer that may be helpful in the development of Faith, and that is the fact that all who master and apply the other sixteen principles of Philosophy of American Achievement will thus have placed themselves within easy reach of faith. This entire philosophy is one of action. It inspires effort based on definiteness of purpose and this is exactly what is required in the development of Faith.

Principle 5 Go the Extra Mile

Going the extra mile is defined as the habit of rendering more service and better service than one is paid for. Some advantages of doing more than one is paid for are (1) Develops initiative and self-reliance (2) Enables an individual to profit by law of contrast (3) Aids in development of attractiveness of personality (4) Ensures continuous employment (5) Develops a positive mental attitude

Principle 6 Take Initiative

Successful men are known, always as men of action. There can be no action without the exercise of one's initiative. There are two forms of action, namely (1) that which one indulges in from the force of necessity and (2) that which one exercises out of choice, on his own free will. Leadership grows out of the latter. It comes as the result of action in which one engages in response to his own motives and desires.

Was Mr. Carnegie's major motive to make money? No, "his motive always has been that of making men more useful to themselves and to others". He made forty ordinary laborers into millionaires.

Mr. Carnegie, doesn't the question of education enter into one's personal achievement? Isn't it true that the educated man has a better chance of success than the man who lacks education? No. Many men are schooled, but few men are educated. An educated man is one who has learned how to use his mind so that he can get everything he desires without violating the rights of others. Education, comes from experience and use of the mind, and not merely from the acquisition of knowledge. Knowledge is of no value unless and until it is expressed in some form of useful service.

All too often, educated men expect to be paid for that which they know instead of that which they do with their knowledge.

Principle 7 Develop Imagination: Cultivate Creative Vision

A philosopher said, "the imagination is the workshop of man wherein is fashioned the pattern of all his achievement."

All forms of organized effort must be planned through Creative Vision. Creative Vision is but another name for imagination.

Synthetic imagination consists of the act of combining recognized ideas, concepts, plans, facts, and principles in new arrangements. One thinker described imagination as "the workshop of the soul wherein man's hope and desires are made ready for material expression."

Creative Imagination is, "The workshop of the soul" through which man may contact and be guided by Infinite Intelligence. Some men perceive and interpret new ideas never before known to man. Edison used synthetic imagination to invent the incandescent electric lamp and he used creative imagination to invent a machine to produce sound he called the phonograph.

Principle 8 Enforce Self-discipline

The major requirement for individual achievement is the subject of self-discipline. This entire philosophy serves in the main to enable one to develop control over oneself, this being the greatest of all the essentials of success. Self-discipline begins with the mastery of one's thoughts. Without control over thoughts there can be no control over deeds! Let us say, therefore, that self-discipline inspires one to think first and act afterward.

Mr. Carnegie, what is your motive in spending all this time coaching me to organize the philosophy of American Achievement?

Answer: I have obsessive desire to provide the people of America with a safe and dependable philosophy by which they may acquire riches in their highest form; riches which will enable people to relate themselves to one another so that they may find peace of mind and happiness and joy in the responsibilities of life.

Self-discipline is largely a matter of adoption of constructive habits. It calls for a balancing of the emotions of the heart and the reasoning faculty of the head.

Principle 9 Organize Your Thinking: Think Accurately

Organized thinking gives benefit of the knowledge, experience, and education to others, through the medium of Master Mind Alliance and develops habit of accurate analysis.

Success is always the result of an ordered life. An ordered life comes through organized thinking and carefully controlled habits. Success is a habit. Work is thought power put into action. Thinking constructively is an individual responsibility.

Principle 10 Learn From Adversity and Defeat

Two important facts of life stand out boldly! One is the fact that the circumstances of life are such that everyone inevitably is overtaken by defeat. The other is the fact that every adversity carries with it the seed of an equivalent benefit!

One cannot get the most out of material wealth unless he or she earns it.

Two kinds of people never get ahead. Those who do only that which they are told to do and those who will not do what they are told to do.

There are millions of people in an imaginary prison who have been charged with no crime. They are prisoners in their minds, consigned there by their own self-imposed limitations, through the acceptance of poverty, and the acceptance of temporary defeat. It is this sort of prisoner that I hope to release, through the Philosophy of American Achievement.

The rescue must begin by awakening them to the realization of the power of their own minds.

Some causes of failure are inadequate schooling, lack of self-discipline, profanity, speaking before thinking, and covetousness.

Principle 11 Seek Inspiration: Control Your Enthusiasm

Inspiration is the emotions put into action through spiritual power. Inspiration is the beginning of all great achievements of whatever nature!

Enthusiasm is a positive expression of feelings and is contagious. Enthusiasm is a builder of confidence.

Emerson wrote: ''Nothing great was ever achieved without enthusiasm.''

Enthusiasm denotes hope, courage, and belief.

So-called genius is only a man who because of great capacity for enthusiasm, steps up vibrations of his mind until he is enabled to communicate with a source of knowledge not available to him through his facility of reason.

Principle 12 Control Your Attention

Controlled attention is the act of combining all the faculties of the mind and concentrating them upon the attainment. The domineering forces in Carnegie's mind during his life was making and marking steel. Controlled Attention is necessary for success in life.

This principle is an indispensable part of philosophy of individual achievement. It makes the other principles work.

Principle 13 Apply the Golden Rule: Develop Positive Mental Attitude

To get the most from the Golden Rule, it must be combined with the principle of going the extra mile, wherein consists the applied portion of the Golden Rule. The Golden Rule supplies the right mental attitude, while Going the Extra Mile supplies the action feature of this great rule. A combination of the two, gives one the power of attraction which induces friendly cooperation from others as well as opportunities for personal accumulation such as friendship and fortune.

Applying the Golden Rule opens the mind for the guidance of Infinite Intelligence, through faith, attracts the friendly cooperation of others in all human relations, eliminates the desire of something for nothing, makes the rendering of useful service a joy that can be had in no other way, makes one a power for good, enables one to recognize the joys of accepting the truth that every man is, and by right should be, his brother's keeper.

These are no mere options of mine. They are self-evident truths, the soundness of which is known to every person who lives by the Golden Rule as a matter of daily habit.

Principle 14 Cooperate: Inspire Teamwork

For industrial America to be successful, it must operate within the principle of team-work.

Never tear down anything unless you are prepared to build something better in its place. The railroads will lose business to the automobile and airplane if the public is not given useful service. The common error of many people is the failure to recognize facts in time to profit by them. Government and industry must work together as partners.

Principle 15 Budget Your Time and Money

The usual distribution of twenty-four hours per day period is: eight hours for sleep, eight hours for work, and eight hours for recreation (free time).

A man may use his free time for opportunities related to his work. Free time may be called "opportunity time".

If you do not like your present work, use your free time to prepare yourself for something better. A man's work can be a recreation if he does it in a spirit of intense enthusiasm. Free time preparation includes reading, schooling, and formation of friends. No man has a right to live unto himself alone.

Money should be put to work earning more money.

Everyone should be on a budget. Savings habit connotes self-discipline.

Principle 16 Make Health a Habit

The physical body is a "house" which the Creator provided to serve as a dwelling place for the mind. Some aids in maintaining sound physical health are mental attitude, eating habits, relaxation, elimination (waste matter), hope, and avoiding drugs.

Principle 17 Convert Mental Desire into Faith: Use Cosmic Habit-force

It is the master key to the universe.

Cosmic Habit-force converts mental desire into a state of mind known as Faith which inspires one to create definite plans for the attainment of whatever is desired.

Genius is born of hardship. of deprivation, and of toil to surmount difficulties.

The man who says "it can't be done" is usually busy trying to keep out of the way of the man who is doing it.

Suggested Reading

How to Raise Your Own Salary, by Napoleon Hill. Chicago, Illinois: Napoleon Hill Associates, 1953

Grow Rich! With Peace of Mind, by Napoleon Hill. New York: Fawcett Book, 1967

The Master-key to Riches, by Napoleon Hill. New York: Ballantine, 1965

Success through a Positive Mental Attitude, by Napoleon Hill. New York: Pocket Books, 1960

Think and Grow Rich, by Napoleon Hill. New York: Fawcett Crest, 1960

You Can Work Your Own Miracles, by Napoleon Hill New York: Fawcett Book (Random House Publishing), 1971

Character Building the Speeches of Booker T. Washington. Executive Books 2006

Advantages of Poverty Andrew Carnegie Executive Books 2004

The Website of the Napoleon Hill Foundation: www.naphill.org

Welding Made Simple

Within these pages you can discover the lost art of the Columbia Gas "Puddle" welding method. After a mere 24 hour shielded arc welding course, with minimal cost and material, this method can help you pass the AWS D 1.1 Structure Welder Qualification Test. Learn from over 45 years of experience as Bill talks you through this simple, cost-saving technique step-by-step.

About the author: William Ballis, a Professional Engineer, worked for Columbia Gas for 31 years before retiring. Shortly after retiring, his wife, Wanda, lost her battle with kidney failure. During this difficult time, the Lord led Bill to youth and adult prison ministries within the state of Ohio. By reaching out to others and sharing his faith, Bill has been able to work throuth his pain and help bring healing to himself and others.

Bonus: This book also contains an appendix that describes principles that Bill has learned from Napoleon Hill's inspirational writings. Bill has introduced these life-changing ideas to prisoners in Ohio correctional institutions over the past several years.